Specialist Floor Finishes: Design and Installation

D. CATTELL
Principal
Cattell Consultancy Services
Crewe, Cheshire

Blackie
Glasgow and London

Blackie and Son Ltd
Bishopbriggs, Glasgow G64 2NZ
7 Leicester Place, London WC2H 7BP

© 1988 Blackie and Son Ltd
First published 1988

All rights reserved.
No part of this publication may be reproduced,
stored in a retrieval system, or transmitted,
in any form or by any means,
electronic, mechanical, recording or otherwise,
without prior permission of the Publishers.

British Library Cataloguing in Publication Data

Cattell, D.
   Specialist floor finishes: design and installation.
   1. Floors, Concrete   2. Finishes and
   finishing
   I. Title
   693'.52   TH2529.C6
   ISBN 0–216–92251–8

Phototypeset at Thomson Press (India) Limited, New Delhi.
Printed in Great Britain by Bell & Bain (Glasgow) Ltd.

Preface

For as long as concretes have been used as substrates there has been a need for upgrading the immediate working surface, in some instances nominally, in others substantially. This in itself has presented problems because of the inexact nature of the art and the conditions under which upgrading is required to be performed, not to mention the variety of surfaces and products.

The aim of this book is to highlight the best parameters for success and to provide the reader with a comprehensive introduction to the subject. I have no hesitation in parapharasing an old saying: 'The man who never laid a problem floor never laid a floor'. This is not to say that it was ever a deliberate mistake, but that the nature of the work is fraught with problems, and that there exist wide ranging possibilities for error. I have had the additional benefit of hindsight of other people's errors and it is to be hoped that this book will promote a more practical understanding of the difficulties involved in upgrading and finishing concrete substrates.

I am exceedingly grateful to an understanding family, to Alison Hartley and the word processor, and to Ronald Gartside and his draughting board.

David Cattell

Contents

Introduction	1
1 Choice of specification	**3**
1.1 Specialist paint and sealers	4
1.2 Polymer or resin modified cementitious toppings	5
1.3 Synthetic resin toppings	7
1.3.1 Epoxy resin surfacing systems	8
1.3.2 Polyester resin surfacing systems	10
1.3.3 Polyurethane surfacing systems	10
1.3.4 Furane resin surfacing systems	12
1.4 Modular finishes	12
1.4.1 Fully bonded tile floors	13
1.4.2 Floating tile floors	13
1.4.3 Elastically bonded tile floors	14
1.4.4 Choice of tiles	15
1.4.5 Choice of bedding and jointing systems for tiled floors	16
1.5 Anti-slip properties	18
1.5.1 Anti-slip versus ease of cleaning	18
1.5.2 Anti-slip finishes for monolithics	18
1.5.3 Anti-slip tile finishes	19
1.5.4 End user maintenance	20
1.6 Other floor finishes	21
1.6.1 Sheet rubber flooring	21
2 Resin systems	**22**
2.1 Epoxy resins	22
2.1.1 The resin	22
2.1.2 The hardener	23
2.1.3 Resin hardener ratio/rate of cure	23
2.1.4 Resilience	24
2.1.5 Moisture tolerance	24
2.1.6 Chemical proerties	24
2.1.7 Physical properties	25
2.1.8 Skin irritation potential	25
2.1.9 Popular hardener systems for flooring application	26
2.1.10 Principal uses of epoxy resins in flooring applications	26
2.2 Polyester resins	27
2.2.1 The resin	28
2.2.2 Catalysts	28
2.2.3 Rate of cure	29
2.2.4 Moisture tolerance	29
2.2.5 Shrinkage	29
2.2.6 Chemical properties	30
2.2.7 Physical properties	30
2.2.8 Skin irritation potential	30
2.2.9 Principal uses of polyester resins in flooring applications	30

CONTENTS

2.3 Furane resins	31
2.3.1 Chemical properties	31
2.3.2 Physical properties	32
2.4 Polyurethane resins	32
2.4.1 The resin	33
2.4.2 Catalysts	33
2.4.3 Rate of cure	33
2.4.4 Chemical properties	34
2.4.5 Principal uses for polyurethanes in flooring applications	34
2.5 Latex cements and mortars	34
2.5.1 Chemical properties	35
2.5.2 Physical properties	35

3 The powder phase — 36

3.1 Powder, or filler	36
3.2 Speciality fillers	38

4 Membranes — 40

4.1 General definition	40
4.2 Need for membranes	40
4.3 Types of membrane	41
4.3.1 Thin polyethylene	41
4.3.2 Bituminous membranes	41
4.3.3 Synthetic resin membranes	42
4.3.4 Polyurethanes	42
4.3.5 Sheet rubber	43
4.3.6 Plastics	43
4.3.7 Asphaltic membranes	44

5 Substrate design — 45

5.1 New substrates	45
5.2 Drainage	46
5.2.1 Screeds to falls	50
5.2.2 Surface finish of substrates	51
5.3 Movement joints in substrate design	52
5.4 Plinths	53
5.5 Design related to non-cementitious substrates	54
5.6 Preparation of new substrates—concrete	55
5.6.1 Vacuum shotblasting	55
5.6.2 Scabbling	55
5.6.3 Acid etching	55
5.7 Existing substrates	56
5.7.1 Removal of existing finishes	56
5.7.2 Moisture	57
5.7.3 Removal of contamination	57
5.8 Repair of prepared or existing substrates	59
5.8.1 Suitable materials for repair of existing substrates	59
5.9 Existing steel substrates	60

6 Design of finishes — 61

6.1 Product design—screeds	61
6.1.1 Polymer modified cementitious screeds	61
6.1.2 Resin modified cementitious toppings	62
6.1.3 Resin toppings	62
6.2 Installation design—topping	63
6.3 Product design—modular finishes	65
6.3.1 The tile	65

6.3.2 The fixing system	66
6.3.3 Movement joints	66
6.3.4 Membrane	66
6.3.5 The site	66
6.3.6 Tile joints	67
6.4 Installation design—modular finishes	68
6.4.1 Bed thicknesses—joint widths	69
6.4.2 Columns and plinths	71
6.4.3 Channels	72
6.4.4 Gulley outlets	75

7 Movement joints   77

7.1 Position of movement joints	78
7.2 Construction of movement joints	79
7.3 Common materials used for movement joints	81
7.3.1 Polysulphides	81
7.3.2 Modified epoxies/polysulphides	81
7.3.3 Polyurethanes	81
7.3.4 Silicone rubbers	81
7.3.5 Bituminous mastics	82
7.4 Installation of movement joints	82
7.5 Movement joint calculation parameters	83

8 Site requirements   84

8.1 Weather protection	84
8.2 Summary of site requirements	84

9 Installation of specialist floorings   86

9.1 Application of membranes	86
9.1.1 Sheet membranes	86
9.1.2 Application of liquid membranes	89
9.1.3 General	89
9.2 Application of finishes	90
9.2.1 Polymer modified cementitious finishes	90
9.2.2 Resin toppings	91
9.2.3 Application of modular finishes	93

10 In service maintenance of floors   101

10.1 General	101
10.2 Cleaning	102
10.2.1 Steam cleaning	103
10.2.2 Cleaning problems	103

11 Fault finding, problem analysis and solving   105

11.1 Fault finding	105
11.2 Problem analysis and solving—tiles	106
11.2.1 Arching	106
11.2.2 Tile edge failure	106
11.2.3 Tile separation	107
11.2.4 Tiles breaking up	107
11.2.5 Tile staining	107
11.3 Problem analysis and solving—joints	108
11.4 Problem analysis and solving—monolithic finishes	108
11.5 Problem analysis and solving—movement joints	110
11.5.1 Bridging	110
11.5.2 Splitting or adhesion failure	110
11.5.3 General problems	111

12 Case histories — 112
- 12.1 Figures 12.1–12.11 — 112
  - Figures 12.3–12.5 — 113
  - Figure 12.6 — 115
  - Figure 12.7 — 115
  - Figure 12.8 — 117
  - Figure 12.9 — 117
  - Figures 12.10 and 12.11 — 117
  - Figures 12.12 and 12.13 — 117
  - Figures 12.14–12.17 — 118

13 Chemical resistance table — 125

Index — 130

Introduction

In pursuing consultancy practices, it became clear, from contact with civil engineering, architectural and consultancy personnel in a wide range of industries, that there was a need to transfer to them a degree of knowledge about protective finishes for concrete and other minor substrates which:

(a) bordered on the technical without being too confusing or so basic as to insult their intelligence
(b) included a limited degree of polymer science to help in understanding the available materials, their differences and attributes
(c) emphasized the all too often forgotten parameters for achieving success in design, preparation and maintenance of flooring systems, and contained the essential elements of the best trade practices
(d) gave an insight into floor designs, materials and services from the theoretical and practical standpoints and from the benefit of hindsight 'enjoyed' by consultants the world over.

In no way will this publication enable the reader to formulate or apply the products discussed. It should, however, enable a better understanding of them and avoid many of the pitfalls experienced in specifying and using flooring systems.

To the end user. I hope you will appreciate that the floor is the most important part of your building. Within reason, you reap what you sow in the financial importance you attach to your floor installation and its after care. The writer accepts also that there are circumstances where early failure has occurred on expensive floor systems. It is hoped that this publication will assist in preventing some of the errors leading to such problems, generally associated with poor application techniques or inadequate preparation.

To the architect and consultant. Please understand that substrates are a vital part of the topping system and must be designed with the topping in mind. Realization that from time to time aesthetics must give way to practicality and fitness for purpose is often an afterthought when it should be a forethought.

To the main contractor. Everybody is concerned with programme, believe it or not, but speed should never be achieved at the expense of quality, particularly as the floor topping is, for the client, the most important part of the building you are constructing. Nobody can guarantee the client's everlasting gratitude for an excellent floor; he probably will not realize he has one. On the other hand, ignore the floor finishes at your peril, the problem will go on and on forever.

To the clerk of works. The successful application of specialist finishes can be as dependent on site conditions as on the specialist contractor you must work with.

Give the flooring contractor space and freedom from other trades until his job is complete and then protect the finish from subsequent operations.

To the contractor. The pressure is great to take the order from the competition, but do not make the mistake of reducing the specification or skimp on the preparation, as it will cost more in the long run. Buy your clients this book so that they may appreciate some of your problems.

To everybody. The floor is an underestimated element in building construction until it goes wrong. The floor is essentially an industrial road, pavement, store, drain, discharger and container of industrial effluents and corrosive fluids, and a protection against chemical and mechanical attack. It is the one part of the building which cannot be avoided; you must walk on it, you must sit on it, truck, build, or support on it, store, spill, drain or contain on it, lift from it, drop or throw on it. The foregoing is nothing new, but how many people appreciate it? Not many. This situation changes immediately someone slips on it or trips on it. Just wait until it leaks—the whole world recognizes a floor with problems.

1 Choice of specification

The choice of and preparation of a specification for flooring systems to protect substrates from chemical attack or provide hygienic finishes is an exercise all too often couched in financial considerations.

Consider the following problems associated with failure of floor finishes due to mechanical damage, abuse, and as a result of inadequate specification:

(a) manifestation of major structural failures, in extreme cases impossible to remedy
(b) harbouring of bacteria in hygienic installations
(c) the need to remove equipment to reinstate or repair floor areas
(d) disruption of production in part or in whole
(e) loss of access to production areas.

And yet despite the importance of floors in industrial installations, specifications for floor finishes generally reflect short-term budgets rather than long-term stability, inviting early failure. Financial considerations for floor finishes should be confined to comparing bids for the ideal specification, compiled with the end use in mind.

There are many specific considerations to be taken into account when deriving a suitable specification for a given situation, not least the following:

(a) the location of the structure to be protected, its method of construction and condition if it is an existing substrate
(b) the composition, concentration and temperatures of all fluids or solids which may be in contact with the floor surface as a result of the inherent processes; if these vary from area to area it should be clearly stated and designated areas clearly marked, to allow selection of materials appropriate to each area
(c) loading characteristics of the substrate and their ability to withstand additional superimposed loads from toppings, tiles, brick or composite surfacings
(d) details of pedestrian and wheeled traffic, including wheel loadings and the nature of wheel contact surface
(e) details of isolated activities such as drum handling and storage and the loadings involved, and the areas to which they apply
(f) the type of floor drainage contemplated e.g. outlets or drainage channels; these may also be part of the contractor's recommendations in the specification along with the details of falls recommended

(g) the nature, concentration and temperatures of all cleaning agents which will be used on the floor, including any materials used for cleaning equipment and pipework since these are usually discharged on to the floor or preferably into channels
(h) details of proposed changes in plant operation involving reagents or conditions not previously notified.

Comprehensive information from the end user in these respects will give the specifier the maximum information with which to derive an appropriate specification for a durable finish.

Having established the parameters for selection of finish there are many systems to choose from and within the systems different alternatives. The floor finishes fall into the following categories, and are considered in this order as systems, then discussed in more detail.

(a) Paint and sealers
(b) Polymer or resin modified cement toppings including terrazzo tiles
(c) Resin toppings
(d) Modular finishes.

1.1 Specialist paint and sealers

Specialist paints and sealers have no place as true corrosion resistant finishes in flooring applications, but serve well where upgrading of new or existing concrete or steel substrates is necessary or desirable for other operating reasons such as dust control. On concrete, sealers can increase surface hardness and abrasion resistance, reduce water absorption and dusting and can generally provide an aesthetic finish to concrete when required, but with limited durability to other than foot traffic. Sealers cannot be expected to improve physical or mechanical properties of substandard concretes by anything more than a nominal amount.

Preparation of the concrete is necessary to remove dust, laitance and general contamination.

This option for substrate protection must only be regarded as a maintenance coating. It has a limited life and is subject to a high degree of wear and tear. The degree of maintenance required for paints and sealers can to some extent be reduced by selecting them only for areas where traffic is light or where pneumatic wheels only are in use.

Paints and sealers must be designed to penetrate the surface to reinforce and be reinforced by the concrete surface, rather than to lie on the surface, which invariably leads to flaking. Penetration is related to viscosity, therefore solvent diluents are often used to achieve this. Industries sensitive to solvent vapours and odours because of a risk of taint or explosion should consider non-solvented systems.

Acrylic, natural and synthetic latices are not considered suitable as sealers,

CHOICE OF SPECIFICATION

Table 1.1 Summary of sealers for specific applications

Sealer	Remarks
Silicates—	sodium and potassium (also known as water glass). Solvent free, suitable for non dusting applications. Dry installations only.
Water miscible epoxy resins	Solvent free, suitable for non dusting duties and for reducing water penetration. Improved abrasion resistance, can be pigmented.
Epoxy resins, polyurethane resins, polyester resins	Generally solvent diluted, suitable for non dusting and for reducing water penetration and absorption. Impved resistance to chemical spillage, but beware—one scratch, one hairline crack in the substrate and problems will occur if the system is intended for chemical duty. Do not be tempted by cost.

although as admixtures to cementitious materials they produce a surface with all the attributes of a good sealed surface. They are therefore considered as a system in their own right.

There are numerous proprietary water-based sealers and upgraders on the market, mainly based on silicates, which have good anti-dusting properties but are mostly water-degradable over a period of time. If solvented materials are not practical and resistance to water of the sealer is required, water miscible epoxy resin systems offer excellent durability. A summary of sealers for specific applications is given in Table 1.1.

1.2 Polymer or resin modified cementitious toppings

Modification of cementitious toppings and concrete in the flooring industry can be likened to the alchemists' attempts to produce gold from base metals. The results have produced a variety of so-called alloys all of which improve a specific property of cementitious materials but none which offer an adequate range of properties to avoid the need for alternative non-cementitious materials as discussed in this publication.

Polymer modified systems involve the replacement of gauging water in conventional mortars and screeds with a low viscosity aqueous polymer emulsion based on materials such as acrylic, styrene butadiene, natural rubber, and vinylidene chloride copolymers. Unlike resin systems, polymeric additives as discussed here for floor finishes have little or no cohesive properties and without the cement content are unsuitable for floor finishes. Resin modified systems incorporate complete self curing resin systems and an aqueous system to hydrate the cement phase. These resins are generally epoxy, polyester and polyurethane.

Both systems improve cementitious compounds by a combination of:

(*a*) reducing water content, thus also reducing ultimate porosity and impermeability

(b) coating cement particles thereby improving resistance to dilute acids and alkalis and improving water absorption
(c) filling interstices of cement, sand and aggregate mixes and reducing water absorption and impermeability as a result
(d) forming a dense surface layer under the effects of trowelling, improving all properties including abrasion and anti-dusting.

Modified cementitious systems are laid at thicknesses of between 6 mm and 20 mm, and normally at 10 mm to 12 mm to achieve a balance between performance and cost. A wide variety of formulations exist, from basic systems, literally replacing water in conventional cement and building sand/aggregates, through to high performance prepacked systems incorporating washed, dried and graded aggregates ranging from silica sand through basalt and silicon carbide.

Uses can extend across the whole spectrum of floor finishes except those where chemical spillages predominate. Modified cementitious systems have expansion coefficients relatively close to those of concrete substrates (degrees C for degrees C) and therefore find considerable use in those areas such as cold rooms for meats and foodstuffs, where their additional benefits of lower water absorption, impermeability and hygiene are an obvious benefit, and temperature changes are gradual or controlled.

Non-dusting surfaces and an ability to produce coloured finishes are important factors in selection of these materials in warehouses and food preparation areas. The most advantageous areas of use for modified cementitious systems are where low chemical spillage occurs but where mechanical damage is prevalent. Thickness for thickness they are good value for money.

No modification system, it must be stressed, will completely protect the cement content from attack by acids and alkalis. Resistance will be enhanced only as long as the cement particle remains protected. Ageing of the resins or polymers, or sometimes purely abrasion, will eventually permit or precipitate degradation of the cement under these conditions.

Cementitious systems offering most resistance to acid and alkali attack are based on modification with polyurethane resin. Temperature resistance with some systems is claimed as high as $120°C$ for short periods, based on the resistance and resilience of the resin and the similarity of coefficients of expansion with conventional concrete. It is worth bearing in mind, however, that even materials with identical expansion coefficients expand at different rates if temperatures are different as they would be on a floor subject to occasional hot spillages. Surface temperatures will always be higher than substrate temperatures in these situations, but it stands to reason that systems with coefficient of expansion close to that of the substrate will perform better than those with considerable variance.

Comparison of the properties of modified cementitious toppings does not readily produce a table which clearly identifies individual systems in a meaningful order of importance other than the two groups, mentioned originally:

(a) *Polymer modified*: acrylic, SBR, natural and vinylidene chloride latices, offering improved non dusting, water resistance, resilience and abrasion resistance. SBR and natural latices are more resilient and abrasion resistant. However, there is limited resistance to oils, greases and fats, although this is less applicable to acrylic and vinylidene chloride systems.
(b) *Resin modified*: epoxy, polyester, polyurethane offering much improved chemical resistance and most of the properties of polymers but with a considerable price premium, which in certain instances bears comparison with resin toppings.

1.3 Synthetic resin toppings

Synthetic resin toppings comprise the synthetic resin and appropriate catalyst or hardener system with a graded washed, dried aggregate or powder, containing no acid soluble matter. For this reason they offer a much improved chemical resistance over resin modified cementitious systems. It is necessary to point out that while in this section the benefits of resin toppings will be highlighted, all resin topping systems have limitations; the chemical resistance requirements of industries such as dairies, breweries, abattoirs, pharmaceutical installations and hygiene industries in general are well within the scope of resin systems. Where solvents and concentrated chemicals are used, however, there are limitations.

It would be prudent to remember that when applied as 5–6 mm finishes, mechanical damage can seriously impair protective properties of such toppings. In the industries mentioned, attack on the substrates following damage would be slow, and while of obvious concern, adequate warning of problems is given to attentive end users. In the chemical industries, where mechanical damage is more prevalent and spillages more aggressive, loss of protection would be extremely quick and damaging, and in these situations tile or engineering brick with impervious membranes should not be far from the specifier's mind as the appropriate specification for long-term protection.

The most common synthetic resins used are epoxy resin, polyester resin and polyurethanes.

A range of products can be formulated, from the 2 mm surface self-levelling low viscosity systems, through the standard 5–8 mm toppings, to heavy duty systems up to 15 mm thick. Anti-slip surfaces can be provided by surface dressings containing angular aggregates and these are discussed later. Synthetic resin toppings have specific advantages in that they can be formulated to resist a wide range of chemical environments and possess properties of fast setting, quick maturity and excellent physical properties.

Broadly speaking, the chemical resistance of any system emanates from the resin itself, information on which will be obtained from the tables. Physical properties are derived by a combination of the inherent properties of the resin and by the aggregate formulation and its ratio to the resin.

All of the resins mentioned have advantages and limitations as classes of synthetic resins, and it must be made clear that their inherent properties can be infinitely varied by the manufacturer, type of resin and type of hardener or catalyst (discussed in more detail in Chapter 2 on resin systems). For the purposes of the following discussion, many generalizations have to be made for the sake of clarity.

In evaluating which synthetic resin system to consider for a specific duty, one must consider:

(*a*) the chemical conditions
(*b*) physical requirements
(*c*) temperature of operation and of cleaning
(*d*) anti-slip requirements
(*e*) cleaning requirements.

It is fair to say that the first consideration in derivation of a specification is the chemical condition. However, reading a chemical resistance table of the synthetic resins, one is likely to be specifying furane resins or polyester resins rather than the more common epoxy resin and polyurethanes.

Unfortunately, furane resins and polyester resins, superior in chemical resistance as they are, have undesirable physical attributes that limit their use as chemically resistant floor toppings (but not as mortars for tiles as discussed later), and will only be used when it is not possible to use epoxy or polyurethane. Clearly, selection revolves around a series of parameters, not least the ability satisfactorily to manufacture and apply the protective system to a standard suited to the duty.

1.3.1 *Epoxy resin surfacing systems*

The use of epoxy resin surfacing systems over the past 30 years has been one of the major advances in substrate protection. Unfortunately, overspecification, poor formulation and poor contracting has occurred frequently, as contractors and formulators have jumped on the bandwagon of the panacea for all flooring ills. Epoxy resin toppings, in the main, however, have an excellent reputation for tolerance of civil engineering conditions and practices provided the essential parameters for installation are met.

A good epoxy resin surfacing system comprises four elements: resin, hardener, powder and substrate. All are covered separately in this book. Failure to understand the importance of any of these in relation to the end use would lead to disaster. The number of systems available, however, makes it necessary in this section to generalize when considering these products.

The thickness of the available systems range from 2 mm to 12 mm. Thicknesses below this range can be considered as paint coatings. Above 12 mm the cost implications are too onerous, approaching or exceeding those for tile specifications.

In the range 2–4 mm, the applications are generally of resin rich formulations, termed 'surface self-levelling systems', or 'self-smoothing systems', producing a smooth, dense surface suitable for light to medium traffic situations. While the high resin content makes these thin systems no less expensive than the more substantial thicknesses, properly laid they have a pleasing aesthetic appearance and an inert surface even when lightly scratched, making them suitable for hygiene applications.

Higher resin contents produce higher coefficients of expansion, and the higher the operating temperature, the lower thickness the coating should be, (not necessarily applicable to trowelled systems). Temperatures above 80° C should be avoided, irrespective of claims to the contrary. For these minimal thicknesses, substrate design and preparation is critical and can add to the overall cost. Surface self levelling systems reflect the substrate undulations and are not, as inferred by the popular misnomer, 'self levelling' other than their own surface. If they were truly self levelling, draining from peaks or high points would seriously weaken performance, so a degree of thixotropy is required to prevent this.

Application of thin epoxy systems is covered in the section on applications (9.2.2), but generally involves the use of serrated trowels, rollers or squeegees to achieve the correct thickness for the application. Spiked rollers are frequently used for achieving air release by encouraging surface flow and a reduction in surface tension.

In the range 4–8 mm, filled epoxy systems are known in the industry as 'toppings'.

Ideal filler to resin ratios for toppings are between 5:1 and 7:1. Below these figures, cost is high and coefficients of expansion border on the unacceptable. Higher filler ratios require less fines to be achievable and the resulting product does not have the desirable density to perform adequately without sealing. Open texture toppings also have low adhesion values due to poor substrate contact. Application of sealer coats to epoxy resin toppings or other resin toppings is an acceptable practice provided its purpose is not to compensate for inadequate density and open texture. Resin toppings of this nature will lose integrity when scarred or worn and any claimed benefit of high filler ratio toppings can be discounted on overall performance.

Similarly, higher filler ratios achieved by diluented resins having lower viscosity will result in lower chemical resistance.

Whereas epoxy resin toppings thicker than 8 mm are practical, cost considerations are too onerous and tile systems may be contemplated on a value for money basis.

1.3.2 Polyester resin surfacing systems

Polyester resin systems are available in forms similar to those of the epoxy resin systems. First glances at the chemical resistance tables indicate that polyester resins have a wider ranging resistance than epoxy resins, particularly where oxidizing chemicals and solvents are concerned. Unfortunately, converting the resin into an acceptable flooring system has met with both opposition and practical difficulty.

The opposition has come as a result of the unavoidable presence of styrene odours with their very low threshold of smell, which have proved obnoxious particularly in the food processing industries where anything threatening to taint is unacceptable. This has somewhat limited the market to chemical industries.

Practical problems relate to three typical properties of polyester: an exothermic reaction on curing, shrinkage and air inhibition.

Exotherm can to some extent be controlled in filled topping systems as the fillers increase bulk, but working pot life is shorter than that of epoxies.

Shrinkage is a problem which can only be mitigated by careful selection and grading of aggregate, and by accurate proportioning of catalysts to cause to react the maximum amount of styrene. Shrinkage is a result of a combination of initial expansion on curing due to exotherm which creates shrinking stresses on cooling, and gradual loss of styrene not fully reacted.

Air inhibition, which is lack of complete surface polymerization, occurs as a result of contact with the air. The only method of overcoming this is to incorporate solutions of wax in styrene which bloom to the surface during cure to form a barrier. These wax solutions have only a minimal effect on highly filled systems, which need to be overcoated with a waxed top coat. Primer systems cannot contain wax solutions otherwise adhesion of toppings will be affected. For this reason, overcoating is carried out on wet primer and the system cures from the top.

It would not be prudent to rule out the use of polyester systems in the chemical industry where solvents or oxidizing chemicals are prevalent. Conventionally catalysed systems incorporating peroxides are limited in thickness of application to around 5 mm by the need to avoid exotherms. Modified catalyst systems can be laid up to 25 mm thick but at considerable cost.

Substrate requirements are identical to those of epoxy resin surfacing systems.

1.3.3 Polyurethane surfacing systems

As a synthetic resin surfacing system and compared to epoxy resins, polyurethane is still in the early stages of development. There were tremendous strides over the years 1976–86 and there is no doubt that the competition to epoxy will continue to strengthen as technology improves.

Polyurethanes have additional benefits as flooring systems. Such resins are inherently more flexible and resilient and therefore have the capability of absorbing expansion to a greater degree with the obvious improvement in service temperature limits. Over the whole spectrum, chemical resistance is equivalent to that of epoxy resins and the lower viscosity of the resin results in dense toppings readily consolidated. The resilience of the finished system is also an advantage if substrates should develop fine surface cracks. Expansion joints can be formed in chemically similar materials having excellent adhesion, good extension and identical chemical resistance.

Polyurethanes are not without disadvantage, particularly in their sensitivity to moisture which reacts with the hardener/catalyst to liberate excessive carbon dioxide gas which results in blistering. Substrates must therefore be carefully prepared and moisture levels determined. These are generally required to be below 5% total moisture as tested by portable electrical instruments.

Priming of the substrate serves two purposes—it reduces contact with inherent moisture and it seals the concrete, preventing air rising during setting which can also cause blistering of the typically fluid systems. Porous concretes may require multiple priming coats to achieve an effective seal.

Carbon dioxide given off in the catalysis process can sometimes be visible on the surface as small inclusions, as the gas is trapped at that point where release of surface tension is no longer possible. These inclusions have no effect on performance being of a closed cell nature, although for the aesthetic quality application techniques should minimize their occurrence.

With most polyurethane systems, surface finish is smooth and glossy. Therefore, should anti-slip properties be required, a surface dressing is necessary that contains a suitable fine aggregate, which should be rounded where ease of cleaning is important.

The more flexible polyurethanes are less resistant to scratching than epoxy systems; harder versions lose a degree of flexibility but retain an inherent resilience that epoxies do not possess to the same degree. Polyurethanes exist in two main types as flooring materials, as follows:

(a) Moisture curing polyurethanes are solvent-based low viscosity translucent-to-clear resins, the light stable versions of which are eminently suitable for producing light coloured and multicoloured floor finishes. As moisture curing systems, they are required to be laid in multiple layers to allow solvents to evaporate and moisture to access. Applied too thickly, skinning will occur which traps solvent and precludes full cure.

Be aware that solvent odours can be a problem in sensitive areas.

Moisture curing polyurethanes are predominately used in light duty pedestrian areas. Thicknesses are low, generally well below 2 mm and use for chemical spillage or process areas is not recommended.

(b) Isocyanate-cured polyurethanes rely on chemical cross-linking to achieve cure and do not contain solvents. Thicker sections are, therefore, possible up to 8 mm, offering a more substantial protection comparable with other resin toppings, temperature resistance depends on the system chosen but can be up to 120° C; the flexible systems have little or no coefficient of expansion.

1.3.4 *Furane resin surfacing systems*

Furane resins find only an extremely limited use as a floor topping despite excellent chemical resistance. The predominant reason is that to all intents and purposes furanes produce only black finishes and they can be bonded to concrete only by intermediate coats of bitumen which serve to protect the acid catalyst in the resin from the alkalinity of the concrete. Adhesion is therefore limited and certainly not adequate to permit trafficking or thermal stresses. Its predominant use is in severe chemical environments in bund areas where traffic does not exist.

1.4 Modular finishes

Technically, the next step up from resin toppings are modular finishes, ranging from thin tiles 10 mm thick through all thicknesses up to 75 mm thick. Determination of the required thickness is made by considering traffic weight and density and levels of mechanical damage anticipated.

Modular tile or pavior systems offer enhanced chemical resistance as a result of the inherent chemical resistance of the tiles themselves combined with the ability to use a wider range of chemical resistant resin jointing systems than are available for use as floor toppings. Modular finishes can be laid as architecturally complementing finishes, and to a greater variety of colours than resin toppings.

Tile floor systems cannot be regarded as impervious in the short term and certainly not in the long term. Thermal stresses, attrition, vibration and even long-term permanent growth of tiles will lead to or eventually promote hairline fracture of the jointing system or tiles themselves. For this reason, membranes are an essential feature of a tiled floor on suspended floors where water leakages can be a problem and mandatory directly beneath the finish on all floors where strong chemical spillage occurs.

In choosing a tile specification it is important to realize that there are a number of methods by which tiles can be laid to suit the end use, and from time to time the wrong decision in this respect is made because of lack of appreciation of the differences. Furthermore the performance of the tiling in flooring applications with respect to ultimate chemical resistance depends more on the system of laying and jointing used than the type or thickness of the tile, provided the tiles are of good quality and below 3% water absorption which permits classification as vitrified.

CHOICE OF SPECIFICATION 13

This rule is not applicable in tank lining situations where continued contact with chemicals demands tiles with very stringent property specifications, and which subject is outside the scope of this book.

The various methods of laying can be applied to the whole thickness range. It is the physical requirements of the floor which determine the thickness used and the chemical resistance requirements which determine the method of laying.

The various methods are as follows and will be further detailed in application of finishes (section 9.2) and design of finishes (Chapter 6): fully bonded tile floors, floating floors, and elastically bonded floors.

1.4.1 *Fully bonded tile floors*

When a fully bonded tile floor is laid, the tiles are adhered directly to the substrate or screed by means of a bed of cement, polymer modified cement or by acid-resistant cements where necessary. The method should only be used on mature substrates or screeds which are sound without random joints, cracking or repairs. There is no opportunity to isolate these systems from the substrate or screed and therefore any movement will be reflected in the finish.

Maximum adhesion must be achieved, particularly when thin beds are used, as the finish in an unbonded condition will have insufficient weight to remain stable unless minimum 50 mm paviors are used. Bonding is achieved by use of neat cement slurries, 1:1 slurries of sand–cement, proprietary bonding agents or, in the case of acid-resistant cements, with a suitable primer. All of these bonding systems are enhanced by substrate preparation such as vacuum shotblasting or scabbling.

Semi-dry beds are quite suited to bonded systems but due care must be taken to ensure the substrate is sufficiently damped to avoid moisture being drawn out of the semi-dry bed, precluding adequate hydration. Where thin bed systems are used, whether cementitiously bedded or by resin, it is imperative that the substrate or screed is true and level (unless to falls), otherwise the bed may need to be considerably increased to compensate.

Acid-resistant jointing systems compatible with cementitious beds can be used to good effect to improve resistance to penetration by spillages, which is important in hygiene industries. Conventional membranes cannot generally be incorporated beneath bonded tile floors (unless already under an existing screed) and therefore with an exception discussed later, these constructions are unsuitable for strong chemical duties. Bonded tile floors are predominately used in ground floor situations where traffic tends to be heaviest and where water leakage through the finish is not of a critical nature.

1.4.2 *Floating tile floors*

Floating tile floors are completely isolated from the substrate and are mostly used in situations where new immature concrete is involved or where the condition of the existing substrate is such that movement, flexing or cracking may be anticipated. To be stable, the minimum thickness of tile and bed should be 50 mm and where the cement bed itself exceeds 75 mm a coarse aggregate should be included.

Floating floors are commonly used in chemical resisting floor situations, with the membrane itself acting as a separating layer. In these situations, as cementitious beds cannot be used and acid-resistant mortars are required to be laid thinly, tiles or paviors of at least 50 mm must be contemplated to be laid directly on to the membrane.

Where cementitious beds are used in floating floors, the separating layer is commonly two layers of polythene. Substrate surfaces must be smooth, and typically of wood float finish if the slip potential is not to be impaired by irregularities with subsequent stressing of the finish.

1.4.3 *Elastically bonded tile floors*

The system of elastically bonding tile floors, developed over the 10 years up to 1986, is not as universally available as the other systems are, because it is based on special materials and technology available from contractors specializing in corrosion-resistant applications. It cannot go without mention, however, as it combines the best properties of both bonded and floating floors and permits the full thickness range of any suitable chemical-resistant tile to be used in fully chemical-resistant applications.

Elastically bonded tile floors incorporate an elastomeric polyurethane membrane system approximately 2.0 mm thick which is bonded to the concrete slab or screed laid to the appropriate falls.

The tile finish from 12 mm to 50 mm can then be bonded (subject to appropriate use of materials technology) directly to the membrane using fully bed and joint methods with selected acid-resistant cements. Selection parameters therefore only revolve around the required mechanical strength of the finish and chemical resistance limits of the membrane and mortar.

The elasticity of the membrane is adequate to retain integrity under normal substrate movements, and to absorb the differential coefficients of expansion between the tile and the substrate, but the rules relating to expansion joints must still be applied to the tile finish. The system retains the solid feel and impact strength of bonded floors with the added advantage of an energy absorbing layer beneath the tile.

An incidental technical advantage of bonding to the membrane is that liquors penetrating the floor finish as a result of mechanical damage cannot travel under the tile finish and thereby escape neutralization or harbour bacterial growth. In the event of membrane damage, leakage would occur at the problem area making it readily identifiable.

Figures 1.1(a), (b), and (c) show the three main methods of affixing modular finishes. In Figure 1.1(a) tiles are laid integrally with the screed bonded to the

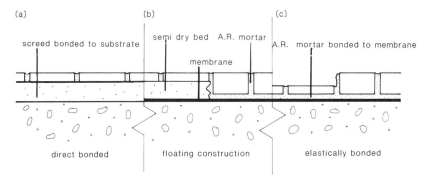

Figure 1.1 Three main methods of affixing modular finishes: (a) direct bonded; (b) floating construction; (c) elastically bonded

substrate or the tiles are bonded to a screed previously laid and bonded to the substrate. In (b), membrane is loose laid or cannot be bonded to, and therefore acts as a separating layer. In (c), membrane is bonded to substrate or screed and tiles are bonded to membrane with chemical resistant mortar.

1.4.4 *Choice of tiles*

Everybody will recognize that the choice of tiles available for use in flooring is considerable in design, colour and thickness. It is the aesthetic possibilities combined with durability that make tiles so popular as a floor finish.

The most suitable floor tiles fall into one of two classes: vitrified tiles and fully vitrified tiles. Vitrified tiles and paviors have a water absorption in accordance with BS 1286 of below 3% (latest figure) and fully vitrified tiles have a water absorption of 0.3% or less. In most applications for flooring systems, selection of tiles that complied with requirements for vitrified tiles would be more than adequate. This category also includes paviors which are essentially thicker heavy duty tiles up to 50 mm thick (in instances up to 75 mm thick).

Because all applications, even those designed to be the same, are different in use, there is no hard and fast rule which will allow selection of a specific thickness of tile or pavior for a specific duty with certainty. No matter what thickness of tile is selected, in most installations there is one area, however small, where the specification could probably have been improved. Tile manufacturers will be able to identify the most suitable product in their range on experience in similar installations, and most end users will have already established an experience, good or bad. Typical broad limits for traffic for tile finishes are as follows.

Tiles 10–15 mm: will withstand pedestrian traffic and wheeled pedestrian traffic excluding steel wheeled, but are not recommended to be laid directly on a membrane unless utilizing the elastically bonded method.

Tiles/pavior 18–30 mm: are suitable for uses as for 10–15 mm tiles plus most vehicular traffic excluding steel wheeled; medium impacts i.e. hoses, small tools and so on. But direct adherence to a membrane is not recommended unless an elastically bonded method is used.

Tiles/paviors 50 mm and above: are generally suitable for most applications. They can be fully bedded and jointed directly over any membrane system. For this reason, paviors are the preferred system for most chemical resistant applications.

1.4.5 *Choice of bedding and jointing systems for tiled floors*

It will have become obvious from the foregoing that there are alternative methods of fixing tile systems and alternative resin materials for jointing. To some extent the choice of jointing resin will influence the method of fixing and vice versa and therefore to complete this chapter it is necessary to include some explanation in this respect.

The least complicated requirements to be satisfied from the options point of view are those demanding full chemical resistance properties from the finish. The chemical conditions dictate the resin system or systems, and the appropriate choice is then used to fully bed and joint the tile or pavior, directly on to the membrane.

The other factors to be considered are to ensure that the resin jointing selected is compatible with the membrane and will not in itself be detrimental to it, and also that the finish selected is of adequate thickness to be stable.

Resin bedding and jointing systems are essentially thin bed systems because

(a) slump of the bedding material can occur as a result of the nature of the resin and the need to maintain an adequately high resin content for chemical resistance
(b) cost is high and therefore it is beneficial to minimize quantities
(c) certain resins are subject to exothermic reactions and the lower the mass the fewer problems occur in this respect.

When this method of bedding and jointing is used, it will be clear that the substrate will be laid to falls before application of the finish or membrane.

In the hygiene industries tiled finishes are commonly laid on a semi-dry bed which incorporates the fall required. However unless a cement joint is acceptable, which in most installations is unlikely, a suitable jointing material must be subsequently used. Resins used for jointing only must be compatible with the cement bed to be effective, for instance furane resin cements are acid catalysed and contact with the alkaline cement neutralizes catalysis and results in a partly set interface. Whereas the interface may be acceptable on a 50 mm pavior when only a small proportion of depth is affected, on a 15 mm tile the proportion is unacceptable. It is recommended that, irrespective of thickness, furane mortar joints are not used where the possibility of contact with cement exists. Epoxy resin mortars are widely used with cementitious beds, polyester resins to a lesser extent.

A resume of the common resin mortars follows. More detailed information is contained in the specific section under resin systems.

Epoxy resin mortars. Epoxy resin mortars are widely used for bedding and jointing tile and paviors in the hygiene industries such as dairies, breweries, pharmaceuticals and general food and meat processing. Most epoxy resins have low installation odour (except phenolic amine hardened), a good tolerance of moisture during setting and water washability (subject to type) which is important with light coloured decorative tile systems. Epoxy resins are compatible with cementitious bedding systems. Chemical resistance is good for the mentioned applications but resistance to oxidizing chemicals such as nitric acid and sodium hypochlorites is limited.

Polyester resin mortar. Polyester resins are only preferred to epoxy resin in areas susceptible to strong chemical attack, particularly by the oxidizing chemicals and solvents to which epoxy resins have limited resistance. Polyester resin mortars in general are not tolerant of moisture before setting and therefore have limited uses in maintenance situations and on moist cement beds. Styrene odours from polyester resins are objectionable where taint is likely. Use of polyester resins, where essential from a chemical point of view, therefore will require more stringent attention to installation parameters.

Furane resin mortars. Furane resins are the most widely used cement in chemical environments. They have excellent resistance to all acids (except oxidizing) alkalis and solvents. Furane resin is not tolerant of water before setting and its dark colour will stain light tiles. Furane is an excellent fully bedding and jointing system and although it does have an odour it is not normally regarded as objectionable. In contact with cement bedding, catalysation is impaired, therefore where no membrane is included concrete screeds must be coated with an isolating primer. Bitumen paints are commonly used for this purpose, but they are not regarded as true bonding agents and promote only minimal adhesion; therefore thin tile systems fully bedded and jointed in furane are not recommended, but could be possibly used in foot traffic areas.

Silicate mortars. Widely used in tank linings for high concentration high temperature acid resistant linings, silicate mortars based on sodium or potassium silicate liquids have limited use in flooring applications because of total lack of resistance to alkalis, poor resistance to water and a porosity of about 10%. They are resistant to all concentrations of all acids except those containing fluorine. They are therefore little used in flooring applications except in bund areas in concentrated acid storage areas.

1.5 Anti-slip properties

1.5.1 *Anti-slip versus ease of cleaning*

Other than complete failure of a floor finish, nothing raises the passions of end users more than the topics of anti-slip and ease of cleaning. The two are closely connected and for this reason it is appropriate to include this aspect of floor finishes within the section discussing the available systems, and to point out that correct selection of footwear will go a long way toward safety on floors with a tendency to be slippery owing to the nature of product deposited.

First we must dispel the misnomer of 'non-slip floors'. Practical non-slip floors do not exist in the sense that they remain so under all operating conditions. Anti-slip is a far better choice of adjective and does not exaggerate an almost unachievable property. Anti-slip surfaces, in reality, offer a reduced chance of loss of traction by pedestrian or vehicular traffic than plain floors. Their effectiveness will be reduced by failure to remove contaminants and frequently their very nature makes cleaning difficult. The end users looking for both anti-slip and easy clean properties will invariably end up with a compromise and would do well to decide in advance which property is the more necessary and communicate it to the designer or contractor.

1.5.2 *Anti-slip finishes for monolithics*

In monolithic finishes (floor finishes that are continuously laid and bonded to the substrate), high filler loadings can give inherent surface profiling of a fine nature which is little better than a smooth finish in other than liquid-only situations. Aggregates such as bauxite, flint and corundum may be incorporated into the filler and be partially exposed on trowelling to break up the smooth surface. Surface profiling with this method is limited in view of the necessity to trowel the surfaces smooth and to avoid depressions around the aggregate which would retain dirt. Included (as opposed to superimposed) anti-slip aggregates such as these do show improved wet slip resistance over plain surfaces but would not be satisfactory where solids are the problem.

One alternative method of achieving an anti-slip property is to superimpose a scatter of materials as previously described after the initial trowelling of the surface. This can only be done as laying proceeds, for obviously the topping must not be set. This process leaves craters around the aggregate and therefore subsequent sealing of the surface with a resin rich coat after setting of the main coat is essential to fill them. Because the aggregate is anchored into the topping itself, the life of the anti-slip surface is relatively good, provided the aggregate has been selected for its durability.

An anti-slip surface can also be achieved by incorporation of varying sizes of coarse sand or aggregate into a sealer coat which is first trowelled over the cured floor and then rolled with a paint roller to produce an even texture.

CHOICE OF SPECIFICATION

In heavy traffic areas, these latter anti-slip coatings can be less than permanent, particularly if large size aggregate is used, as the particles can be broken out. Very coarse sand or fine corundum-filled resins heavily dressed over the surfaces sealed have better longevity provided they are well bonded. Additional aggregate-free resin coats may be necessary to reduce the surface profile if more cleanable surfaces are required. When considering any anti-slip but cleanable finish in monolithics, small samples should not be relied on to give a clear picture of the finish. Distribution of anti-slip finishes can be extremely variable and trial areas with the various types of finish are best evaluated before final decisions are made.

1.5.3 Anti-slip tile finishes

Anti-slip tile finishes are in the main more durable and easier to clean than monolithics. Prefabrication enables a wide variety of profile shapes to be produced to a regular pattern and intensity. Samples are more representative of ultimate finishes and can be checked well in advance of application for the appropriate properties, and many coefficient of friction calculations are already available for a wide range of profiling and contact materials of which a selection are given in Table 1.2.

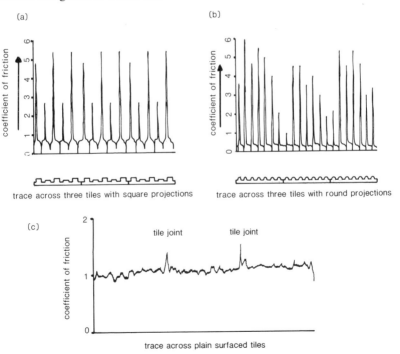

Figure 1.2 Effect of tile joints on coefficient of friction and influence of tile profile; trace across three tiles with: (a) square projections; (b) round projections; (c) plain surface

Tile finishes also have inherent anti-slip properties from the joints in addition to any profiles, or rough surfaces in the module. The traces in Figure 1.2 (courtesy BCRA) show the effect of tile joints on coefficient of friction and the significant contribution of differing profiles to anti-slip properties.

Installers may well charge a slight premium for laying anti-slip tiles because cleaning off the cementitious bed or resin joint is more difficult initially. It is not unknown for masking compounds to be used before jointing to ease cleaning before hand over. This does not necessarily point to difficulties in cleaning in use, as the removal of cement and resin jointing is far more difficult than material such as food residues. The masking compound will also minimize staining of tiles from dark resins.

1.5.4 End user maintenance

Anti-slip surfaces require that the end user maintains their surface efficiency by regular cleaning and general good housekeeping. It is water, product spilt on the floor, and general accumulation of loose material which will create the slip potential. Fats, soaps and food products, chemicals such as sodium hydroxide, petrols and oils all promote slipperiness or clog anti-slip finishes to such an extent that the property is non-existent and non-effective. For these reasons, an end user purchasing an anti-slip floor is likely to be wasting money and risk insurance claims unless an effective cleaning regime is carried out.

The trace reproduced in Figure 1.3 (courtesy BCRA) gives an indication of the loss of coefficient of friction due to surface deposits which clog the anti-slip profiles.

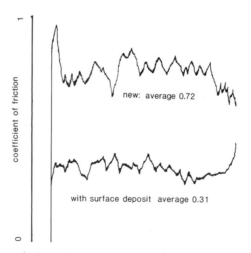

Figure 1.3 Loss of coefficient of friction due to surface deposits which clog anti-slip profiles

Table 1.2 Friction values for anti-slip surface tiles (courtesy BCRA)*

	Hard rubber slider			Leather slider		
	Dry(s)	Wet(s)	Wet(f)	Dry(s)	Wet(s)	Wet(f)
Small studs	0.77	0.73	0.72	0.83	0.81	0.75
Rough texture	0.77	0.66	0.76	0.85	0.79	0.72
Lozenge Projection	0.92	0.94	0.70	0.90	0.96	0.90
Square block plus small studs	0.84	0.70	0.66	0.90	0.94	0.84
Round blocks	1.62	1.58	1.44	1.60	>2	1.05
Triangle blocks	0.76	0.74	0.72	0.84	0.81	0.71
Irregular studs	0.77	0.83	0.90	1.01	0.95	1.02
Part rough texture	0.73	0.70	0.70	0.82	0.75	0.73

*Friction values can alter with the rate of skid and therefore figures are given for two speeds of test: (s) = slow $0.3\,\mathrm{cm\,s^{-1}}$ and (f) fast $7.4\,\mathrm{cm\,s^{-1}}$.

The judicious selection of footwear where anti-slip finishes are required is complementary to its effectiveness. It may have been apparent from the traces that leather produces better anti-slip properties, but under normal circumstances leather soles are not practical in permanently wet or chemical spillage areas. Good quality rubber soled safety shoes, boots or wellingtons with profiled soles are normally recommended.

1.6 Other floor finishes

1.6.1 Sheet rubber flooring

Vulcanized rubber sheeting, in the main anchored to substrates by latex screeds, has become increasingly popular in non-chemical environments, specifically in areas where heavy objects are dropped such as kegging plants and loading and unloading areas where the noise absorbing properties of the rubber can be put to good use. While chemical resistance is good with some limitation on grease and solvent resistance, such floors are not recommended for areas of aggressive chemical spillage due to difficulty of jointing adjacent sheets to an adequate standard to maintain liquor impermeability.

2 Resin systems

2.1 Epoxy resins

Epoxy resins are so widely used as coatings, toppings, sealers and primers, they are deserving of more detailed attention than most resin systems.

The technology of the epoxy resin, hardener and/or catalyst systems is a science in its own right, such is the variety of systems available and the complex nature of their chemistry. In this section we deal entirely with those aspects related to its use in whole or in part, as a protective finish for flooring applications.

Epoxy resins are without doubt the most extensively used resin binders for chemical resistant specialist composition floors. It also has to be said that the widespread use has also brought problems of over-specification, poor formulation and poor application.

The specific benefits of epoxy resins systems are:

(a) versatility of application
(b) excellent adhesion to a wide range of substrates
(c) resistance to dilute acids and alkalis
(d) adequate degree of solvent resistance except to ketones and chlorinated solvents
(e) excellent physical properties
(f) low shrinkage on curing unless containing solvents or water
(g) low installation odour.

What is always not realized is that epoxy resins per se do not offer the optimum of properties listed. Specific requirements are achieved by careful selection of hardener systems and fillers, and in most instances they are compromises in the chemical and physical property sense. Selection is based on the predominant requirements of the finished floor.

The final choice of system i.e. resin, hardener and filler must also be practical and therefore capable of satisfactory application under site ambient conditions. The resin and hardener are dealt with here, the fillers or powder phases, the formulation of which applies equally to all resin systems, are rightfully given attention in a separate chapter.

2.1.1 *The resin*

Epoxy resins are available in both solid and liquid forms, in a range of viscosities. The liquid forms are invariably used for flooring applications. Reactive and non-reactive diluents are frequently used to provide a comprehensive range of viscosities. Reduction of viscosity provides better flow properties particularly in surface self levelling systems (otherwise know as self levelling or self smoothing) and permits higher filler loadings. Diluents are essentially cost-effective, but detract from general physical and chemical properties (particularly solvent diluented), but they have a place where these are not critical parameters.

In food industry applications, solvent free and diluent free systems are normally essential to avoid tainting of in-manufacture food stuffs.

The epoxy resin base is, in effect, the only common denominator in completed formulations. Equivalent epoxy resin specifications vary little from manufacturer to manufacturer. The basic method of comparing resin with resin is by measurement of the epoxy equivalent weight, a measurement of the reactivity per unit weight. Standard undiluted epoxy resins have epoxy equivalent weights of about 180–200. Diluted systems have lower epoxy equivalent weights.

2.1.2 *The hardener*

Epoxy resins may be hardened by crosslinking with other chemicals or by addition of catalysts which promote polymerization, or by a combination of both. Flooring systems in the main utilize hardeners and therefore the emphasis will be on these.

Hardener systems are far more varied than resins, both in the chemical sense and in the number of suppliers. The hardeners impart the most critical aspects of the finished systems; the same base resin can be given a wide range of properties (and lack of them) by varying the hardener system.

Essentially, the nature of the hardener affects:

(*a*) resin hardener ratio
(*b*) rate of cure
(*c*) resilience
(*d*) moisture tolerance
(*e*) chemical properties
(*f*) physical properties
(*g*) skin irritation potential.

2.1.3 *Resin hardener ratio/rate of cure*

Early hardener systems based on primary and tertiary amines were very reactive and required relatively small quantities per unit resin i.e. up to 12 parts per 100. It was therefore necessary for mixing to be very thorough if consistent curing was to be achieved, unless the hardener had some form of catalytic effect which promoted cure even if not fully mixed. Unfortunately, the primary and tertiary amines also proved to be severe skin irritants and be relatively volatile. Fast rates of cure also lead to short pot lives and exotherm, therefore handling proved difficult for the typically large batches required for flooring applications. Present day hardeners (as discussed later) based on polyamidoamines and cycloaliphatic amines utilize higher ratios, 30–60 parts per 100 resin, and permit large volume mixings with reduced tendency to exotherm.

2.1.4 *Resilience*

Resilience is influenced by the hardener, in that systems can be formulated to be from hard and unyielding, to soft and flexible. Certain variation is possible by altering mixing ratios, but this is not recommended and the desired state of cure is generally acquired by choice of hardener type.

Flexibility can be achieved through flexible epoxy resins or flexible hardeners, or a combination of the two. The chemical and physical properties of resins flexibilized are inferior to conventional systems.

2.1.5 *Moisture tolerance*

Many hardeners and, therefore, compounded resins are susceptible to the presence of high humidity in the atmosphere or in the substrate. Problems arising from moisture contamination may manifest themselves as poor adhesion, poor state of cure, and poor intercoat adhesion. It has to be accepted that for flooring formulations, some degree of moisture tolerance is essential due to the nature of the surfaces to be treated and building site and in-manufacture environments.

It is possible with certain hardener systems to incorporate water into the mix as a cost-effective diluent for damp surface primers. Chemical resistance is considerably affected and these combinations are best not considered for chemical duty other than as primers where they are not exposed. Water miscible hardeners can be very useful for incorporation into epoxy bound cementitious mixes for concrete repair and maintenance.

Water dispersible primers are generally milky in colour when first mixed. The milkiness disappears as the water evaporates and curing takes place. In all instances before overcoating, the primer should have become clear.

2.1.6 *Chemical properties*

It is an unfortunate fact that the hardener systems offering best chemical and solvent resistance are not conducive to widespread use as resin flooring systems, due to lack of moisture tolerance, odour, poor colour stability, toxicity and so on. They may, however, be considered in special instances as bedding and jointing systems for acid-resistant tiles. Even in these applications, water tolerable systems are preferred for ease of cleaning after laying.

With the extensive range of hardeners available it is difficult to be specific

Table 2.1 Resistance indications of epoxy systems

Chemical	Indication
Up to 25% H_2SO_4	Resistant
Up to 5% HNO_3	Resistant
Up to 35% HCl	Resistant
Up to 25% NaOH	Resistant
Up to saturated ammonia	Resistant
Naphtha	Resistant
Paraffin	Resistant
Toluene	Limited
Methyl ethyl ketone	Limited

about the suitable maximum concentrations of acid and alkalis to which resistance is provided, or even about resistance to varying pH levels, because resistance or otherwise is more a question of the type of acid or alkali rather than a generalization of strength as indicated by the hydrogen ion concentrations or pH.

A chemical resistance table is incorporated within the book. However, to give the subheading some meaning, in Table 2.1 are given abbreviated resistance indications for various classes of acid, alkali and solvents for ambient temperature spillages as occurring on floors. For containment or higher temperatures, specialist advice must be sought.

It cannot be overemphasized that because of the variation within hardener systems and so on, there is no substitute for consulting a specialist contractor or supplier about the resistance limits of his particular systems.

2.1.7 Physical properties

The physical properties are varied by hardener probably to a lesser degree than chemical resistance. The formulation of the filler phase also has a major influence on compressive, flexural and tensile strengths, and therefore the formulation, if required to arrive at specific minimum physical properties, should take this into account.

A typical range of physical properties should be:

Tensile strength	10–20 N/mm^2
Compressive strength	60–100 N/mm^2
Flexural strength	30–40 N/mm^2
Heat distortion temperature	70–80° C

2.1.8 Skin irritation potential

This aspect of the health and safety consideration of epoxy resin formulations cannot be ignored. Application of epoxy resins as mortars, or as toppings, invariably involves handling of the final mixture and inevitable contact with the skin on some part of the body, even when properly clad. Epoxy resins themselves are known to be skin sensitizers, but hardeners can generally be more aggressive. Present day systems due to the pressures of the Health and Safety at Work Act are safer than the primary and tertiary amines formerly in common use.

In all instances when handling epoxy resins, barrier cream should be used in combination with hand and forearm protection, and specialist resin removing creams used for removal.

2.1.9 *Popular hardener systems for flooring application*

Polyamidoamines (polyamides). Polyamidoamine hardeners are widely used for toppings and mortars and offer adequate acid/alkalis resistance, but only fair resistance to solvents.

Colour stability varies with type, and has a tendency to be darker than cycloaliphatics, therefore polyamides are less used in light-coloured toppings. Water miscible and tolerable systems are normally compounded from polyamides.

Cure times are prolonged at temperatures below 10°C and to avoid retardation they are generally modified with blends of faster curing amines. The resilient properties of polyamide cured epoxide systems is very good.

Cycloaliphatic polyamides. Hardeners with cycloaliphatic polyamides are generally of lighter colour than polyamidoamines, and have good light stability which benefits their use in coloured toppings and mortars. Chemical resistance is better than polyamides particularly to solvents, but generally inferior to aromatic amines. Resistance to atmospheric moisture is good, but cycloaliphatic polyamides are not recommended for water miscible or tolerant systems. Pot life and exotherm is very satisfactory for flooring applications.

Aromatic amines. Aromatic amines are typical of the systems offering the best chemical and solvent resistant properties. Specific formulations can be made to produce extended and fast cure times but unfortunately the negative aspects of this class of hardener limit its use to areas where its use is essential, and not just desirable.

These negative aspects are related to its odour which can taint foodstuffs, its dark initial colour and poor colour stability. Problems related to skin sensitization are greater than with other hardener systems.

2.1.10 *Principal uses of epoxy resins in flooring applications*

Acid/alkali resistant mortar. Epoxide resins are extensively used as fully bed and joint, and as joint only mortars. The main use is in food, beverage and other hygiene industries, where the chemical resistance requirement is well within the product's capability.

Selected epoxy systems can be applied under the damp conditions associated with such plants as dairies and breweries, and for new installations its compatibility with cement beds and good adhesion to ceramics offers an ideal solution to the problem of sealing tile joints without resorting to expensive (but the best) fully bedded and jointed systems.

The fact that epoxy mortars can be formulated to be water washable before setting is also an advantage over other resin systems when utilizing decorative tiles where aesthetics are paramount, and staining of tiles is detrimental and undesirable. Water tolerance facilitates the use of water for cleaning excess jointing materials from the surface of the tile by a washing action which also gives a reasonably flush joint. Systems washed by solvents tend to be absorbed into the surface of the tile, resulting in stains which predominate when the floor is dry, and of course there is the solvent odour which is undesirable in hygiene industries.

Epoxy resin to powder ratios for fully bed and joint mortars are frequently 1:4 or 1:4.5, giving a plasticity suitable for 'buttering' and for ensuring joints are in compression by floating one tile or brick up against the other.

Gunned or squeegee grouted systems, which are used for pointing only, are more often 1:3 resin to powder ratio. This method is commonly used when tiles are bedded in sand cement or polymer beds as a method for filling the joints.

Epoxy resin toppings. Without doubt, epoxy resins are the most widely applied resin topping system. Average thicknesses are 5–6 mm for trowelled toppings and 2–3 mm for surface self-levelling systems.

Chemical resistance is good to excellent. However, it is not recommended that topping systems as a whole be selected for use as predominant protection of substrates from chemical attack.

Epoxy resin toppings are well suited to the requirements of dairies, breweries, abattoirs and hygiene industries as a whole. The use of surface dressings serves to provide smooth dense surfaces and if anti-slip aggregates are included with the dressing they are suitable for wet applications.

Tolerance to a wide range of ambient conditions is good but many installations have caused problems because epoxy resin has been discounted as requiring the specific attention to detail that all resin toppings must have.

2.2 Polyester resins

Polyester resins, while bettering epoxy resin over the full spectrum of chemical resistance, have never been able to match the versatility of epoxy resins in application as toppings although use as mortars is common. Within this range are also included vinyl esters which offer improved chemical resistance particularly to oxidizing chemicals, but whose application parameters and physical nature are to all intents and purposes identical to polyesters.

While offering improved resistance in particular to oxidizing chemicals (including nitric acid) bleaches and solvents, polyester resins as floor topping systems have been beset over the years with difficulties related to shrinkage, lack of resilience and installation odour problems. They are mainly used as specialist floor toppings for particularly aggressive environments where preferred flooring systems are unsuitable.

Polyester resins for toppings and mortars can be formulated as two or three pack systems which essentially combine the three elements of resin, catalyst and filler.

2.2.1 *The resin*

Unsaturated polyester resins are essentially liquids comprising a solution of a polyester in styrene monomer. The styrene enables the resin to cure to a solid by crosslinking the molecular chains of the polyester under the influence of an accelerator and/or catalyst. There are three main types of resin available: isophthalic polyester resins, bisphenol polyester resins, and vinyl ester resins. All are of lower viscosity than epoxies; other systems available are specialist materials not suited to flooring applications.

Of the materials mentioned, isophthalics are most widely used, offering adequate chemical resistance with relatively uniform curing properties at ambient temperatures. Bisphenol and vinyl ester resins with excellent chemical resistance are considerably more expensive and more sensitive to environmental fluctuations applying to industrial flooring. All liquid polyester resins exhibit the characteristic styrene odour, which having an extremely low threshold of smell, has been one of the practical limitations on development and application of polyester resin floor toppings particularly in foodstuffs installations, dairies, breweries, chocolate factories and other sensitive installations which, chemical industry apart, probably utilize the majority of resin toppings.

For most ambient temperature installation, polyester resins are pre-accelerated at source with cobalt soaps or tertiary amines.

2.2.2 *Catalysts*

Polyester resins are cured by addition of a catalyst which promotes the crosslinking of the polyester and the styrene. The rate of setting in pre-accelerated resins is controlled by the reactivity of the catalyst added. Quantities of catalyst should not be altered from the specified percentage, typically 2%. In particularly cold situations, additional accelerator may be incorporated or a more reactive accelerator/catalyst combination utilized to compensate.

Catalysts consist in the main of organic peroxides such as MEK peroxide, benzoyl peroxide and cyclohexanone peroxide in the form of liquids, powders or pastes.

Most reliable results are achieved by addition of the liquid catalyst to resins immediately before addition of the fillers. Two pack systems utilize powder catalyst dispersed in the filler, which while convenient, can create problems if the filler content requires varying due to ambient conditions. It will be understood that at 2% additions, catalysts must be thoroughly dispersed if complete crosslinking is to be achieved and those filler systems containing catalyst must be added as complete units to avoid problems with settlement or lack of dispersion of catalysts within the filler.

Modifications of catalyst systems with isocyanates have proved to impart improved properties with respect to solvent resistance of polyester resins compared to peroxides alone.

2.2.3 *Rate of cure*

Rate of cure is clearly a feature of accelerator and catalyst percentages and temperature. Exotherms are always generated by polyester resins and it is therefore important to maintain mixed materials in low volumes i.e. spread out rather than left in mixing containers.

Polyester resins have a much faster rate of cure than epoxies. However, this can be disadvantageous in topping and mortar applications, and formulations should aim for a pot life of 40–60 min. Excessive pot life can lead to loss of styrene and incomplete crosslinking; too short a pot life will create application problems and considerable waste.

2.2.4 *Moisture tolerance*

Tolerance of polyester resins to moisture in toppings or to atmospheric moisture is not good and this will reflect in the site and substrate requirements laid down for this type of material.

2.2.5 *Shrinkage*

Another limiting factor in the use of polyester resin for toppings has been its tendency to produce systems with relatively high shrinkage due to styrene loss, particularly in those systems which for reasons of site ambient conditions or inadequate mixing have not been fully crosslinked. At worst, the shrinkage manifests itself as cracking, but normally is exhibited as curling at perimeters. Unrestricted shrinkage of polyester toppings can be between 0.2% and 0.4% which relates to between 12 mm and 24 mm in a 6 m bay. This figure is mitigated by adhesion to the substrate, but it will be understood that considerable stresses will be built up within the coating and will be restrained only as long as the adhesion levels are maintained, emphasizing the need for excellent adhesion.

30   SPECIALIST FLOOR FINISHES FOR CONCRETE

Table 2.2 Polyester resins: chemical properties

Chemical	Isophthalic	Bisphenol
Up to 40% H_2SO_4	Resistant	Resistant
70% H_2SO_4	Limited	"
Up to 40% HNO_3	Resistant	"
Up to 35% HCl	Resistant	"
Up to 25% NaOH	Resistant	"
Up to saturated ammonia	Resistant	"
Naphtha	Resistant	"
Paraffin	Resistant	"
Toluene	Resistant	"
MEK	Resistant	"
Chlorinated solvents	Limited	"
NaOCl (sodium hypochlorite)	Resistant	"

2.2.6 *Chemical properties*

As already indicated, chemical resistance tables are included with the publication. Table 2.2 is given as a resume for a range of chemicals for ambient temperature spillages as occurring on floors. For containment, specialist advice must be sought. The information is given as a general guide. Variations due to resin type, degree of crosslinking and so on are not allowed for.

2.2.7 *Physical properties*

Formulation of fillers, and resin to filler ratios has a major influence on physical properties and a typical range would be:

Tensile	15–25 N/mm^2
Compressive	80–120 N/mm^2
Flexural	25–35 N/mm^2

2.2.8 *Skin irritation potential*

As with all resins personal hygiene is of paramount importance. Both resin and catalysts are detrimental to skin, and protective clothing and barrier creams must be used. Under no circumstances should solvents be used to clean skin. Proprietary resin removing creams are available for removal from hands and other skin.

2.2.9 *Principal uses of polyester resins in flooring applications*

Acid/alkali resistant mortars. Polyester resins are more widely used as mortars than toppings, specifically in tiled areas subject to spillages of nitric acid, oxidizing acids in general, sodium hypochlorite, chlorine and solvents. Polyester resin mortars cannot be water washed and therefore extreme care must be taken in cleaning joints with solvent otherwise staining of the tile will result.

Lower resin viscosities permit resin powder ratios of 1:5 to produce adequate plasticity. Gunning of joints is limited due to exotherm of large mass mixes which if occurring in equipment can be very costly.

Toppings. Situations in which polyester resin toppings would be used in preference to epoxies are all related to chemical resistance, particularly those where oxidizing situations prevail and where tiled finishes are inappropriate or undesirable through cost or due to the need for a cleanable or decontaminable floor surface with specific chemical resistance. Most mechanical and physical attributes can be matched by epoxies.

2.3 Furane resins

Furane resins are primarily used as acid resistant mortars. Lack of cohesive strength and a predominantly dark brown to black colour place severe limitations on other uses in flooring situations as do their adverse reaction with alkaline surfaces such as concrete.

Furane resins are polycondensation products of furfuryl alcohol frequently modified with furfuraldehyde. The formulated resin is catalysed by use of acidic powder catalysts blended with the fillers. Exothermic reactions take place during catalysation and the nature of the catalyst and its reactivity must be carefully controlled if this is not to impair the mortar's efficiency and make application difficult. Effective catalysation is normally achieved by a blend of fast acting catalyst and latent catalyst. A blanced percentage of fast acting catalyst promotes initial set without exotherm and the latent catalyst achieves complete cure over a period of days.

Because the rate of setting of furane resins is directly related to the degree of acidity, problems of inadequate setting can occur if the resin system is in contact with alkaline surfaces such as concrete. This is one of the prime reasons why furane is limited in topping applications. Where used as a mortar for tiling, if not laid directly over a membrane, a priming system frequently bituminous in nature is used to isolate the resin from the concrete. This method does not constitute a means of bonding to concrete, and is not recommended for thin tile systems which may separate and arch.

2.3.1 *Chemical properties*

Furane resins are the most utilized bedding and jointing system in the chemical industry. Resistance to acids and alkalis is excellent although certain formulations have slightly inferior resistance to strong alkali.

Resistance to solvents is second to none, temperature resistance is over $140°$ C, but furane resin has limitations to oxidizing acids such as nitric acid

Table 2.3 Furane resins, chemical resistance

Chemical	Indication
Up to concentrated H_2SO_4	Good
Up to 80% H_2SO_4	Resistant
Up to 5% HNO_3	Limited
Up to 35% HCl	Resistant
Up to concentrated NaOH	Resistant
Chlorinated solvents	Resistant
Toluene	Resistant
NaOCl (sodium hypochlorite)	Not resistant

and oxidizing chemicals such as sodium hypochlorite. In these situations polyesters are utilized. A brief resume of chemical resistance is shown in Table 2.3.

2.3.2 Physical properties

Tensile strength	8–16 N/mm^2
Compressive strength	30–50 N/mm^2
Flexural strength	10–20 N/mm^2

2.4 Polyurethane resins

Chemically, polyurethane resins are a very diverse range of resins, formed by crosslinking a polyol (the resin base) which could be as simple a material as common castor oil and a catalyst, normally an isocyanate such as diphenylmethane- 4, 4'-diisocyanate (MDI).

As with epoxy resins, the technology of polyurethane systems is a science in its own right, further complicated by the wide range of potential resin bases without the standardization of epoxies. For this reason we will again deal with aspects of polyurethanes related to use as a flooring system.

Polyurethanes offer the most threat to the position of epoxy resins as flooring systems, offering equivalent chemical resistance and with the added advantage of an inherent resilience which can be created in epoxy resins only with a loss of chemical resistance. On the negative side, polyurethanes are sensitive to moisture, requiring much closer attention to site conditions, substrate preparation and general application parameters.

Expansion coefficients of sand filled polyurethanes are negligible, permitting use of selected systems at up to 120° C. Cementitious polyurethanes have expansion coefficients akin to concrete and therefore appropriate expansion joints must be included.

2.4.1 *The resin*

The resin base of the polyurethane resins are termed polyols. They are liquids of varying viscosity but normally lower than epoxy resins. As already stated, the polyol can range from basic materials through complicated polymers and therefore the chemistry is variable.

Polyols have negligible odours unless modified with other materials (although priming systems may be solvent carried) and are essentially harmless products suitable for most applications (see catalyst).

2.4.2 *Catalysts*

The most common catalyst system for flooring urethanes is MDI. Unfortunately, MDI is a toxic product which in its raw state must be kept well away from foodstuffs. Under normal application conditions, the combined system is suitable for application in hygiene industries including food preparation areas taking normal precautions. Below 40° C, MDI (subject to grade) is not volatile and is safe to use in well ventilated areas. Polyurethane flooring systems should not be applied to hot surfaces in view of the catalyst volatility. Manufacturers' instructions with respect to health and safety must be clearly followed.

2.4.3 *Rate of cure*

Rate of cure can be varied with system, but pot lives are typically 45–60 mins, which is acceptable, as polyurethanes can be applied relatively quickly. Rate will vary with temperature and exothermic reaction will occur in large volumes. Water contamination and high humidity must be avoided at all costs.

Table 2.4 Chemical properties of polyurethane resins*

Chemical	Indication
Up to 40% H_2SO_4	Resistant
60% H_2SO_4	Satisfactory
Up to 20% HNO_3	Resistant
Up to 20% HCl	Resistant
Up to 50% NaOH	Resistant
Up to saturated ammonia	Resistant
Naphtha	Limited
Paraffin	Resistant
Toluene	Not resistant
MEK	Resistant
Chlorinated solvents	Not resistant
NaOCl	Resistant

*Chemical resistance will vary with type.

2.4.4 *Chemical properties*

Abbreviated chemical resistance for use as a flooring material is given in Table 2.4. For containment seek specialist advice.

2.4.5 *Principal uses for polyurethanes in flooring applications*

Membranes. Liquid polyurethane resin membranes offer considerable advantage over sheet membranes, as they are capable of being laid jointless to steel or concrete substrates. The combination of chemical resistance, flexibility and durability during subsequent topping procedures such as tiling are not matched by other materials.

Toppings. Polyurethane toppings are finding favour in competition with epoxy resins particularly as, with developing technologies, systems are available that range from soft flexible toppings to tough hard toppings capable of withstanding vehicular traffic while retaining resilience. Thickness of application is generally 4–6 mm, although as with all materials systems are available which can be laid up to 25 mm thick.

Polyurethane toppings are widely used in hygiene industries and their generally smooth surface is easily cleaned. Textured surfaces are applied as a surface dressing where anti-slip properties are required.

Mortars. Polyurethane mortars are not commonplace, as the material has a tendency to slumping, and adhesion to tiles without priming is not adequate. The advantages of flexible mortars in traffic situations are dubious for reasons of lack of support. However, it may be possible in the future to formulate systems of adequate supporting ability, while retaining the resilience necessary to eliminate expansion joints.

2.5 Latex cements and mortars

While largely superseded by resin cements and mortars in industrial flooring, latex or latices are worth a mention in a historical context as they were by far the most used bedding and jointing system in dairies, breweries and similar industries for 30 years and were one of the original specialist mortars.

Based normally on a high solids content of 60% natural rubber latex, the mortar is formed by mixing with a blend of high alumina cement and blended graded sands, without addition of any further water. Curing additives to vulcanize the latex are added to the fillers.

The resulting mortar has excellent adhesive properties and retains a high degree of resilience if not flexibility. The use of the high alumina cement gives resistance to dilute organic acids common in the industries mentioned and the mortars could be applied to damp surfaces and were tolerant of high moisture levels. The emergence of moisture tolerant epoxies was not the only factor that has led to replacement of latex cements and mortars; the tendency to utilize stronger and more powerful cleaning agents and detergents required the use of epoxies from a chemical resistance standpoint.

Latex slurries are still used as bonding slurries for tiles bedded on semi-dry screeds and have specific advantages over conventional cement mortar in impermeability, resilience and adhesion. Costs are of course higher.

2.5.1 *Chemical properties*

Latex cements are resistant to dilute organic acids such as those degradation products of milk and beer, and to chemical and physical situations as arising in abattoirs or meat processing chill rooms. It is not recommended in what might be termed chemical environments. Latex cements are not resistant to solvents.

2.5.2 *Physical properties*

Tensile strength	7–8 N/mm^2
Compression strength	10–20 N/mm^2
Flexural strength	18–25 N/mm^2

3 The powder phase

3.1 Powder, or filler

The third phase of most resin cements and the second phase of some is the powder or filler phase comprising silica sand or quartzite, whose importance as an element of a flooring mixture is often underestimated.

Fillers have become an accepted part of civil construction in the form of sands and aggregates and, while most people will have seen the effects of too much or too little sand in a mortar, it is not appreciated what other properties the powder phase imparts such as flow properties, and, conversely, support properties to resin cements. It is important not only in cementitious mortar and toppings but in resin toppings and mortars. In these, the powders are not just extending fillers that cheapen the final mix, although that they do very well, but materials that have other important attributes. Such a filler can act as

Extender and cheapener of cementitious and resin mixes
Reducer of exotherm
Flow modifier
Stiffener
Plasticizer
Carrier of catalyst systems
Improver of physical properties.

Except in those instances where sands are chemically attacked and alternative materials must be used, the filler has little effect on the chemical resistance of a resin or cementitious mixture provided it is not impure. It may increase water absorption but that is regarded as a normal physical change of a filled structure, which cannot be avoided.

The most variable features of sand aggregates are shape and particle size. A third, colour, may be important in light coloured mixes, but the first two play an important part in selection for particular properties such as flow, filler ratio and plasticity.

Sands are classified into 18 categories as detailed in Figure 3.1 (courtesy British Industrial Sand). For resin toppings and mortars, it is not feasible simply to buy bags of sand to formulate a mix even if it is of the correct nature, clean, dry and free from acid solubles. If we consider Figure 3.1 showing the variety of classifications for sand particles, the more rounded the shape, the lower the surface area per unit volume than an angular sand particle of the

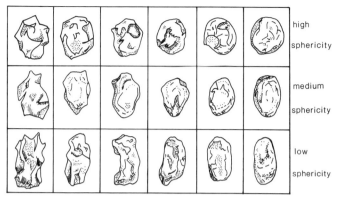

Figure 3.1 Classification of sands

same grading. The lower the surface area the wetter the mix in comparison with the same filler ratio of angular sand, as the coating surface for the resin is smaller and excess resin acts as a lubricant between particles. The rounded sands, because of their shape, particularly moving toward the high sphericity shapes with their smooth surfaces, promote more mobility of the mix and therefore improve flow properties, favouring them for self-levelling formulations, but not for systems where slump is undesirable such as those for vertical surfaces.

Very fine particle sizes, because of the high surface areas per unit volume, lead to low filler ratios by weight and form doughy sticky mixtures. Large particle sizes permit high filler loadings by weight, have poor anti-slump properties, trowel easily but are prone to result in porosity. In successfully arriving at necessary properties for mortars and toppings therefore, it is essential to blend a variety of materials of differing sizes and shapes.

Mortars are required to be smooth and dense, capable of easy buttering, pointing and gunning, but not adhering to tools. They must be plastic enough to spread under hand pressure when positioning tiles, but firm enough to support the tile in the fixed position. As a rule plasticity is much lower than with topping mixtures and resin to filler ratios are in the region of 3 to 5 parts aggregate to one of resin. Mortars require more fines than do toppings, with a proportion of coarser angular aggregates added to reduce stickiness and ease buttering.

Toppings on the other hand are required to be dense but plastic and capable of an easily trowelled tight surface. Ease of trowelling is complemented by round coarse particles, but the balance of fines must be there to fill voids and avoid resin drop out.

As topping fillers are somewhat coarser than mortar fillers, filler ratios are higher at 6–7 parts per part of resin. Actual ratios will depend on the viscosity of individual resins, and it would not be true to say that products with filler ratios outside of these figures were not properly formulated. The formulation and ratio of powder must, however, be such that the performance of the finished product is not in any way impaired.

Fillers for specialist finishes must be clean, dry and acid washed to remove soluble particles and impurities such as iron and acid soluble salts. Suitable sands are sold in this condition and it is not necessary for this to be an operation concerning the purchaser.

Storing and blending must be achieved under conditions that will not adversely affect the product and will maintain its dry condition. As many fillers are intended for use with systems which are moisture sensitive, it is not unusual to incorporate, within the filler, molecular sieves to preferentially absorb normal moisture levels. They are not, it must be stressed, designed to compensate for or reduce excessive moisture levels.

3.2 Speciality fillers

More so in chemical-resistant mortars than in resin toppings (for the reason that resin toppings are not normally used for strong chemicals), it is necessary for certain chemical environments, particularly those involving fluorine compounds, to avoid the use of siliceous fillers which are attacked by these products.

The most common replacements for these conditions are graphite or corundum. Unfortunately, while available in various particle sizes, there is no scope for formulating graphites in the same way as sands. Particle sizes are generally smaller and therefore filler contents by weight are much lower, producing rather sticky mixes. The lubricating effect of the graphite particles produces a more mobile mortar which may result in slumping of the modules. Graphite is used, therefore, in situations where its chemical properties are needed, as application properties of the mortar are then sacrificed.

Graphite or corundum filled mortars are extremely expensive due to the low filler contents and high material cost. This type of filler is used where necessary with epoxy, polyester and furane resins for bedding and jointing tiles or engineering brick. The unsuitability of graphite and carbons in general for toppings is related to colour, cost, reduced physical properties and difficulty of application.

As bulk fillers in specialist mortars and toppings, sands (and graphite where necessary) are used virtually exclusively. Other materials too numerous to mention are, however, added as trowelling aids, anti-slip and abrasion resistant aggregates and conductive materials for anti-static systems, the most common of these being:

(a) *Bauxite*: used as a contrast material in terrazzos and as an anti-slip aggregate in resin toppings

(b) *Calcined plint*: used as a contrast material in terrazzos and as an anti-slip aggregate in resin toppings
(c) *Corundum*: a common material for upgrading polymer and resin toppings, having excellent wearing properties and anti-slip properties but is costly to incorporate
(d) *Granite*: crushed granite is used for terrazzos and for anti-slip aggregates and it is probably the most economically priced product for these purposes
(e) *Pulverized fly ash*: PFA is a lightweight aggregate used for bulking out all manner of topping materials, including resin systems; permeability and water absorption of PFA filled systems is increased and therefore they are normally used for thick section underlays with a suitable topping
(f) *Anti-static aggregates*: aluminium powder, graphite and carbon coated sands are available for utilization in anti-static toppings; however, their incorporation does not guarantee the desired properties and methods of incorporation are a specialist skill.

4 Membranes

4.1 General definition

In the context of this chapter, membranes are intended to imply those materials applied as an impervious layer in liquid or sheet form which are designed to protect the supporting substrate from water penetration or corrosive attack in the event of loss of integrity of the surfacing material due to chemical attack, mechanical damage, thermal stresses or age.

Membrane systems designed to prevent rising damp or hydrostatic pressure are not included except that certain membrane systems may have this property as an incidental additional benefit.

4.2 Need for membranes

Membranes are an essential element of a floor finish if the spillages are corrosive to concrete or steel substrates, or purely if waterproofing of suspended floors is required. In ground floor situations, membranes are included when the spillage is aggressive or when the products of degradation of spillages may become so.

Unless the floor finish is itself a modified membrane, impermeability cannot be assumed. In particular, tile or brick finishes must never be regarded as liquor tight even when new, and monolithic finishes can lose integrity through a variety of reasons throughout their service life.

The position of the membrane within the floor is dictated by various parameters but mainly by the nature of the finish and the nature of the spillage. The most suitable position for a membrane in any installation is of course directly beneath the protective finish, but as monolithic surfacing materials need to be adhered to and supported by an adequate substrate, this is clearly not possible. One of the limiting factors on the use of resin toppings in chemical spillage environments is that membranes cannot be satisfactorily positioned beneath the finish and mechanical damage can lead to failure of the supporting screeds by chemical attack. Similarly, thin tiles or tile and bed systems less than 30 mm thick cannot be used on floating membranes, or in floating finishes, as they will have inadequate weight and depth to resist arching. In these instances, the membrane will be superimposed on the structural slab preferably to falls and screed of adequate thickness applied followed by or laid integrally with the finish.

The limitations of this procedure in chemically corrosive environments can clearly be understood, and while structural substrate units may be protected by the membrane, the finish can hardly be effective if its own supporting screed has been attacked.

With the membrane directly beneath the finish both the finish and the membrane need to lose integrity before failure can occur.

One of the prime requisites of a membrane is an ability to retain resilience and flexibility throughout operating life. The integrity failure of superimposed finishes is normally manifested as cracking, due to mechanical damage and/or thermal stresses, and the membrane should be designed to resist failures of this nature and limit the spread of failure to the finish or topping. In effect, membranes are the protective layer for the substrate and finishes are mainly a means of protecting membranes from the rigours of traffic be it pedestrian or heavy vehicular traffic.

The design of floors to receive membranes should take into account that membranes themselves should be laid to an adequate fall to drains, so that liquors permeating the floor finish are allowed to dissipate and not pool. Laying membranes on a flat floor is a last resort and, even then, the drainage points into which the membrane is terminated must be flush with the general floor level. When luting flanges are provided as membrane terminations and the drain point is higher than the membrane level, weep holes must be provided in the area of the luting flange as indicated and sketched in Chapter 6. All membranes should always be properly terminated at gulleys and perimeters to ensure tanking.

4.3 Types of membrane

4.3.1 *Thin polyethylene (500–1000 gauge)*

This product, commonly available as a separating layer, has no use as a chemical resistant or even waterproofing layer. It is included only to be discounted as contributing to integrity of floor finishes except as a separating layer.

4.3.2 *Bituminous membranes*

Bituminous emulsions are not recommended as efficient membranes. Systems exist, however, comprising bitumen in sheet form supported and strengthened by a layer of thin polyethylene which, properly and carefully laid, can be an efficient waterproofing system.

As with any sheet membrane, jointing is all important and as the system is laid without joint preparation, relying on inherent tack of the products, joint contamination in civil engineering environments cannot be discounted. As with all overlapped seams, installation should start from the lowest point of fall to ensure joints do not face potential liquor flow.

Bitumen based membranes have a high plastic deformation and a poor resistance to cutting, therefore great care must be exercised in avoiding puncture in subsequent laying of screeds or finishes if integrity is not to be impaired. Similarly, the surface of the substrate should not have protuberances which may be forced through the membrane by the weight of the finish. The polyethylene layer precludes bonding of finishes and therefore the minimum thickness of finish must be 30 mm. Bitumen and reinforced bitumen membranes are not recommended for chemical resistant duties except where very dilute chemical situations are involved. Bituminous systems are positively not suitable where any solvents may be present.

4.3.3 *Synthetic resin membranes*

It has already been noted that an important factor in a successful membrane is resilience, elasticity and flexibility and these are certainly valid requirements. There are circumstances, however, where chemical conditions are so severe that resilient systems are completely unsuitable. In this situation, certain synthetic resins, such as polyester and furane, may be used to impregnate glass-fibre mat, producing a reinforced resin membrane with loss of elastic properties over conventional systems but with excellent chemical resistance.

The chemical resistance of the synthetic resin membranes is as listed in the tables, furane being in particular resistant to solvents, but limited in resistance to oxidizing chemicals. Polyester resins bond well to clean dry concrete with laitance removed. However, furane resins are acid catalysed and the alkalinity of concretes will have a retarding effect. It is therefore necessary to incorporate an isolating coating between the substrate and the membrane to prevent contact. Resins such as epoxy and polyester can be used to seal concrete substrates to receive furane resins, but bituminous coatings are commonly used.

4.3.4 *Polyurethanes*

The most versatile synthetic resin in the membrane sense is polyurethane. This polymer has an inherent flexibility and resilience and is applied in liquid form to produce a jointless membrane which can be bonded to steel or concrete substrates.

Polyurethanes are resistant to acids, alkalis, certain solvents, oils and fats. Adhesion to dry concrete is excellent but primers should always be used to seal the concrete surface and improve surface strength. As polyurethanes are applied as liquids, the sealing of the surface is also essential to reduce the effect of air rising from the concrete and to mitigate moisture contamination.

Rather like epoxy resins, the polyurethane chemistry is extensive and complicated and a wide variety of systems are available; particularly varied is the polyol or resin base.

The two main types of polyurethanes used in flooring systems are single pack moisture curing and two pack chemical curing. For use as membranes, the moisture curing systems have limited use, being solvent based and requiring application in numerous thin coats to achieve full and even cure for a suitable thickness. Chemical cure systems can be applied in thicker layers (multiples of 1 mm) provided adequate care is taken to allow gases liberated during the curing reaction to escape.

By far the main disadvantage of polyurethane systems is their sensitivity to moisture. Catalyst systems tend to react with moisture and release excessive carbon dioxide which results in foaming and blistering. Specific attention should be paid to supplier's recommendations on moisture contents in concrete substrates, and to avoidance of contact with moisture and high humidity before full cure is achieved.

4.3.5 *Sheet rubber*

Sheet rubber would appear to have excellent properties as a membrane particularly as there are a variety of polymers available offering a wide range of resistance and elasticity, but with few exceptions, natural rubber, neoprene, nitrile and butyl polymers are not widely used because:

(*a*) costs of conventional sheet rubber, calendered or extruded, are high
(*b*) unvulcanized rubber has a very high plastic deformation and can be readily damaged; uneven loads applied to the subsequent finish may, as a result of the high plastic deformation of the membrane before vulcanization, cause hairline fractures in the finish due to movement and flexing
(*c*) prevulcanized rubbers have excellent elasticity and resilience, but require considerable joint preparation and must be bonded with contact type adhesives which can be difficult under civil construction situations.

A common compromise in the use of sheet rubber is a polymer not indicated previously, polyisobutylene. Polyisobutylene (PIB) is a fully chemically saturated polymer which has good to excellent resistance to acids and alkalis, depending on the grade selected. Vulcanization is not possible or necessary from a chemical resistance point of view but the material does have high plastic deformation, and a lack of elastic properties. These two features are undesirable and great care must be exercised in preparing the substrate and in laying the finish if puncture is to be avoided.

High chemical resistance grades are expensive and cannot be solvent jointed as can the conventional grades. The joint is made by heating the joint interface with a specially adapted hot air welding torch and sealing the joints in a semi-welding technique. Thickness of application is normally 1.5–2 mm.

4.3.6 *Plastics*

Sheet plastic membranes based on polyethylene and polypropylene up to 1.5 mm thick were at one time the premier membrane materials for chemical resistance. Offering excellent chemical resistance with toughness and flexibility to cope with subsequent floor laying procedures, these plastic materials also offered resistance but not impermeability to solvents. This latter feature does not give cause for concern in flooring situations.

Disadvantages of polypropylene and polyethylene membranes are related to installation and, in the main, cost. Installation demands the use of experienced plastic welders, a skill not renowned among installers of flooring systems, and as the sheet sizes are 8 ft × 4 ft or 2 m × 1 m, joints are frequent and obviously critical to a sound installation. Jointing is made difficult by the limited thickness. Electrical testing of the joints can be carried out by utilizing self-adhesive aluminium tape under joints to provide an adequate earthing potential.

Sheet plastic membranes can be laid over damp substrates provided contamination of the joint by moisture does not occur before welding.

The inability to bond plastic sheet membranes securely gives rise to advantage and disadvantage. In a substrate, movement joints can be disregarded in the sense that stresses in the substrate are not transmitted through the membrane to the finish. It must be noted that adequate movement joints must still be provided in the subsequent finishes. The disadvantages are related to the fact that finishes cannot be bonded to the membranes and for that reason:

(a) finishes must have adequate weight to hold down the membrane as it has a buckling tendency due to high coefficients of expansion in changing climatic conditions
(b) finishes must be structural in their own right
(c) liquors penetrating the finish as a result of hairline cracking or mechanical damage and so on are free to pass between the finish and the membrane, making neutralizing difficult and leak detection even more so.

4.3.7 Asphaltic membranes

Asphalt was certainly one of the earliest membrane systems, having good resistance to water penetration, resistance to acids (except strong oxidizing and fatty acids), dilute alkalis and salt solutions. Asphalts have little or no resistance to oils, greases and solvents and in high temperatures are prone to softening and flow, and embrittlement takes place over extended periods. Asphalt has found uses as a combined tanking and floor finish in many situations. However, modern materials for both light duty pedestrian work and heavy duty chemical floor have superseded asphalt with its limited aesthetic properties and predominant installation odours.

Provided asphalt satisfied other selection parameters, its use could be contemplated in areas where its applied thickness could compensate for inadequacies in level.

5 Substrate design

5.1 New substrates

All substrates, irrespective of their nature of construction, must be independently capable of remaining stable without taking into consideration any structural benefits of the proposed finish. Furthermore, particularly in suspended floors, the design should take into account the weight of the finish to be applied. When substrates are steel or timber, deflection must be avoided, as by far the majority of finishes will not tolerate flexing without some loss of integrity. In those finishes that can provide adequate resilience, either by design or by inherent material properties, proper attention must be paid to the manufacturer's limitations, which may vary considerably from product to product and manufacturer to manufacturer.

Existing substrates must of necessity comply with the foregoing requirements. However, certain aspects of the design, if the construction has been made without a topping in mind, may need to be altered or may need a compromise in the finish. In this section we will deal with design for new constructions; existing substrates are dealt with in a separate chapter. It is not intended that this publication delve too deeply into the actual composition of the substrate itself as this is a separate technology and practice covered in parallel publications. It is necessary, however, to point the way to the necessary requirements for substrates, to enable satisfactory floor finishes to be applied to them and to indicate how design considerations with respect to the substrate can benefit the ultimate finish. As a general rule, if the substrates are designed with the following basic parameters in mind, the minimum of problems should be encountered, certainly as far as the suitability of the substrate is concerned.

Briefly, the main points to consider are as follows.

(a) The substrate must have adequate physical properties to withstand operating parameters with respect to loadings, so that undue stresses are not placed on the finish.
(b) Under no circumstances should additives such as water repellents, which may have an effect on the adhesion of subsequent toppings, be utilized in concrete mixes to receive toppings. Shutters for such areas as columns and plinths should not be treated with release oils or paraffinic substances because, if transferred to the concrete surfaces, these products cause serious adhesion problems and are difficult to remove or neutralize.
(c) Movement joints must be included to appropriate standards for the

particular substrate and proposed topping to concentrate stresses at points where they can be dealt with and where they will have minimum detrimental effect on the performance of the finish.

Without doubt, movement joints must be included at perimeters and fixed points as well as in suitable bay sizes.

Consideration must also be given to the positioning of movement joints in the substrate in relation to the most suitable position in the ultimate finish, which, apart from those positions mentioned, is corresponding to the high points in the falls. See Chapter 7 on movement joints.

(d) The surface finish of the substrate must be in accordance with the requirements of the finishing specification as laid down by the contractor, or supplier of the product.

(e) The degree of fall should also be considered in relation to the finishing specification, as well as to the duty involved. Self-levelling finishes should not be laid on steep falls; reference should be made to manufacturers' limits for the appropriate material. Falls must always be adequate to discharge aggressive or high temperature fluids as quickly as practical. This latter situation must be compared with the requirements of the traffic situation and if necessary a compromise made between degree of fall and frequency of channels or outlets.

(f) The substrates should be mature and have completed inherent shrinkage and be typically four weeks old. Moisture contents must comply with the manufacturer's tolerances for the finish, as certain finishes are highly susceptible to moisture contamination during the setting phase, i.e. urethanes. Others such as epoxy are less so, but in high temperature applications water vapour trapped in the substrate in any substantial quantity may cause subsequent blistering, and for this reason, proper assessment with appropriate equipment should be made.

For situations where thin finishes are to be applied, such as resin toppings or tile finishes with thin resin beds, it is the writer's experience that even the best civil contractors, other than on small areas, have difficulty in producing falls in the substrate to adequate surface tolerances or levels, without the incorporation of a screed to falls separate from the main slab. Falls in main slabs are of course a technically satisfactory way of producing the falls, but it must be borne in mind that to correct inadequacies in the falls in the substrate utilizing the resin systems themselves is a costly exercise, and is sometimes not practical, particularly with self-levelling systems.

5.2 Drainage

One of the most important aspects of all floor designs is the adequate and speedy discharge of liquors to drains from the general surfaces, the rate of which is governed by the falls to which the floors are laid. It is not normally

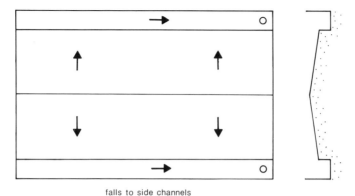

Figure 5.1 Perimeter channel drainage, side channels

appreciated that the degree of fall affects the rate of heat absorption from hot liquors and rate of attack by aggressive liquors. The quicker the liquors move, the less damage they do. Falls should not be too long, particularly on suspended floors, as the thickness of the slab would need to be increased. Long falls involve long paths for liquor flow which should be avoided, and in these instances intermediate channels may be included. If the substrate is to be laid flat and a topping to falls is to be laid, the need for falls on the finished floor affects the levels of the substrate, and the positions of drains, gulleys or channels that pass through the substrate.

Where a thin surface finish is to be laid, the falls must be already incorporated in the substrate, which would comprise slab with screed to falls. With certain tile finishes, generally those laid on a semi-dry bed, falls are built integrally with the floor finish. There are various methods of achieving adequate draining:

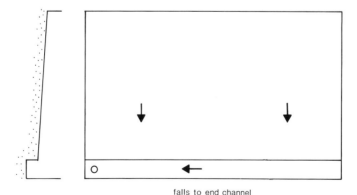

Figure 5.2 Perimeter channel drainage, end channel

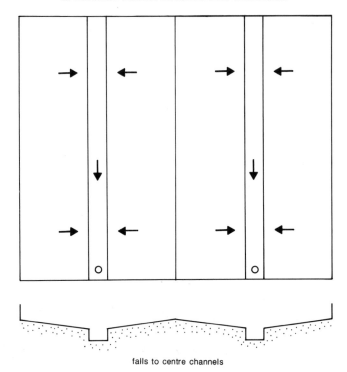

falls to centre channels

Figure 5.3 Centre channel drainage

(a) perimeter channels as depicted in Figures 5.1 and 5.2
(b) centre channels as depicted in Figure 5.3
(c) isolated channels as shown in Figure 5.4
(d) gulley outlets as depicted in Figures 5.5 and 5.6.

The degree of fall and type of drainage selected will be influenced by the nature and degree of spillage and the type and intensity of traffic. Areas

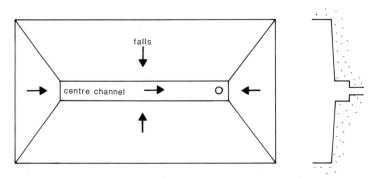

Figure 5.4 Isolated channel drainage

SUBSTRATE DESIGN 49

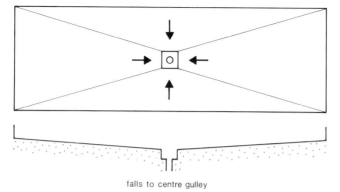

falls to centre gulley

Figure 5.5 Central gulley outlet

beneath plant and equipment may be steeper to remove liquor quickly. However, traffic and pedestrian areas must take into consideration slip resistance and levelness for those purposes.

Suggested minimum falls to achieve adequate drainage would be 1:100. Suggested falls for high temperatures and highly corrosive fluids would be 1:40. Falls greater than this become safety hazards and difficult for traffic, so a compromise for general purposes and safety is 1:80. It is worth repeating that pooling due to inadequately screeded falls in the substrate may reflect in the finish, and be impossible to remove.

Where possible, overflow and discharge from machinery and plant should be piped directly into channels, and the position of such plant should be taken into consideration when designing drainage, even if it is necessary to create small isolated channels for this sole purpose.

Note: If channel gratings are required, rebates will be necessary. The method for forming the rebate will vary with specification. Provision must be made for adequate terminating of the specification at the rebate and various suggestions

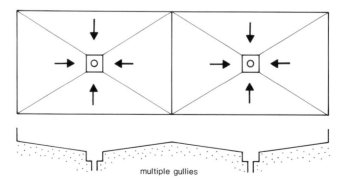

multiple gullies

Figure 5.6 Multiple gulleys

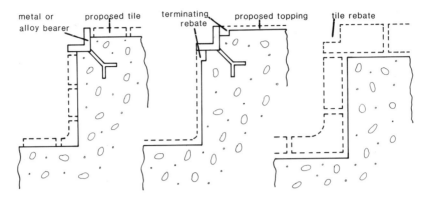

Figure 5.7 Methods of terminating specification when channel grating installed

are sketched. If metal bearer bars are used with monolithic systems, chases must be formed. In all specifications the height of the bearer bar above the substrate must allow for the thickness of the finish as per Figure 5.7.

There are few installations free from columns, plinths, service ducts and general interruptions to the surface and specific advantage can be gained by adequate and proper attention to design considerations; alternatives are discussed later.

5.2.1 Screeds to falls

Having already made it clear that screeds are the most satisfactory way of achieving falls, there are two alternative methods of placing the screed in relation to the main slab

(a) as a non-bonded or isolated system
(b) as a bonded system.

Over many years, both have been put forward as the preferred method and it would be appropriate to discuss the merits of each.

Isolated systems. Isolation is normally achieved by incorporation of a heavy duty polyethylene layer between substrate and screed. Isolation is also achieved as a matter of course if a waterproofing membrane is incorporated in the same position.

The principle behind isolation is to prevent physical influences of the substrate reflecting in the finish whether it be resin toppings or tile finishes. These influences in immature substrates may be shrinkage forces which, particularly with tile finishes bonded to the slab, may create serious problems. In this situation, the substrate shrinks to a far greater extent than the tile finish, placing the tiles in compression and developing shearing forces. The shear forces can be enough to break the bond between the tile and bed or substrate, and because the tiles are in compression, arching occurs which can be broken down by subsequent traffic. Once tiles, even individual ones, become displaced under these conditions, loss of compression takes place and the tile finish quickly breaks up. Isolated screeds should also be considered in suspended floors where variable stresses such as those over structural beams may occur.

Isolated systems permit the construction of falls and crossfalls irrespective of the substrate configuration or position of movement joints. Movement joints must always be incorporated in isolated screeds, at perimeters, fixed points and so on and be followed in subsequent toppings.

Isolated screeds less than 40 mm thick are suitable only for light traffic situations, interior use without high temperatures and not for external situations. Isolated screeds over 40 mm and preferably over 50 mm can be used in most situations and environments, subject to proper laying procedures.

One of the negative factors of isolation of screeds is the tendency to develop a curl on drying. This would affect tiled systems with semi-dry beds to a lesser extent as the integral semi-dry screed is less prone to shrinkage. However, thicknesses of at least 50 mm are recommended.

Where curling occurs it generally manifests itself at a perimeter, which may be a wall or at an expansion joint. In extreme circumstances falls can be affected and under heavy traffic conditions excessive flexing of movement joints can occur with possible failure. Control of water content in isolated screeds is therefore extremely important to control shrinkage and curling.

Bonded systems. Screeds properly bonded to substrates by cement slurries, proprietary bonding agents or by polymer slurries in combination with surface preparation should not develop curling. However, the layout of bays must faithfully follow the substrate, particularly at movement joints, and this may be a limiting factor if existing joints are randomly positioned. Similarly, existing substrates which have been extensively repaired will not be suitable for this method. Substrates must also be fully mature if shrinkage forces are not to reflect in the finish, therefore direct bonding to new screeds is generally out of the question unless it is by way of a flexible membrane system which bonds to both screed and finish and absorbs some of the shrinkage forces. See design of finishes (Chapter 6).

Bonded screeds are more resistant to mechanical shock, transmit less vibration, generally take on a more solid feel than isolated systems, and are capable of taking extremely heavy traffic including in exterior use.

5.2.2 *Surface finish of substrates*

Equally critical to the successful application of a topping or finish is the nature of the surface provided to the new substrate. The needs vary with specification and are briefly as follows.

(*a*) For thin topping systems or thin bed tile systems laid directly on to a substrate or screed to falls and for most membranes, the surface must be free from laitance, dust and be generally sound. Holes in the surface may lead to air pockets and should be filled. Protuberances may puncture a membrane during installation or even at a later stage during service due to the applied load of the finish or subsequent traffic. Wood float or light brush finish generally produces a satisfactory surface for all applications.

(b) Thicker systems or thick bed composite topping systems can be laid on tamped concrete provided a membrane is not required at substrate level (in which case (a) applies).

In all instances, the requirements of the contractor or material supplier with regard to surface finishes are paramount. Failure to consult with the specialist contractor may lead to unnecessary expense in correcting substrates.

5.3 Movement joints in substrate design

'Movement joints' are intended to embrace, as a definition, those joints constructed in elastic or plastic materials designed to absorb expansion, contraction, flexing and vibration within floor finishes. Movement joints in specialist floor finishes are often considered weak points because they break floor continuity and are generally dissimilar materials with differing chemical properties to that of the floor finish. Movement joints and immediately adjacent areas have apparently (and factually) high failure rates for numerous reasons, contributing to the reputation.

Except in flexible and isolated finishes, movement joints in the substrate will always be reflected in the finish and should be kept to a minimum conducive with satisfactory design. Designers should bear in mind that judicious positioning of joints in the substrate will reduce premature failure (i.e. at high points).

Positioning of movement joints is an area where design consideration and practical application must compromise. Perimeter movement joints are least troublesome to traffic and where these alone will suffice, others should not be used. In general, traffic areas should be avoided if possible when placing movement joints.

Perimeter joints should not be placed immediately against the wall unless absolutely necessary as they are difficult to access for preparation. Experience shows that if joints are difficult to access they will not be properly prepared or installed in the finish.

The most appropriate distance from the wall varies from specification to specification. In tile finishes movement joints are best sited one tile plus the cove skirting away from the wall, and in monolithic finishes at least one float width away (otherwise the finished surface on the upstand side of the joint will be less than desirable): Figure 5.8.

Where joints must be included other than at the perimeter, they should be

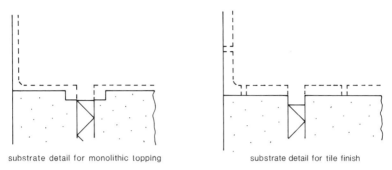

Figure 5.8 Appropriate distance from wall: (left) monolithic topping; (right) tile finish

positioned corresponding to high points in the finish. This will enable specialist contractors to install joints in the finish in a position least susceptible to attack. Note that irrespective of joints in the substrates, the specialist contractor may install joints within the finish independent from, and in addition to those in the substrate.

Movement joints provided in substrate or screeds to directly receive a resin topping must be constructed with true arrises otherwise failure of the topping will occur adjacent to the joint.

5.4 Plinths

The need to place plant and equipment off the floor for access and cleaning or purely to provide a level base in floors with falls generally requires construction of plinths. Except that they interfere with the floor surfaces, particularly when setting out for tile finishes, plinths do not present much of a problem to specialist flooring contractors, but they may require to form movement joints around them in the finish.

As far as the design of the plinths as part of the substrate is concerned, they should not interfere with the general flow of liquors to the channels nor be capable of forming pockets of aggressive liquors. The height of plinths will be dictated by the needs of the installation and whether or not the top is to be capped with a protective covering. Transfer pumps or other units prone to leakage will require protection to the top of the plinth as indicated in chapter 6 on design of finishes.

The consideration given to plinth designs must also relate to supporting columns. Where columns are steel, they should be encased at the bottom with a concrete plinth as indicated below in Figure 5.9.

While flooring openings are one of the least desirable aspects to be designed into a floor, they can sometimes be a necessity. They must never be included without a perimeter upstand, preferably having a height corresponding to the wall perimeter skirting height. If when the final equipment is fitted—and it is presumed that if the need for the opening has been established at design stage

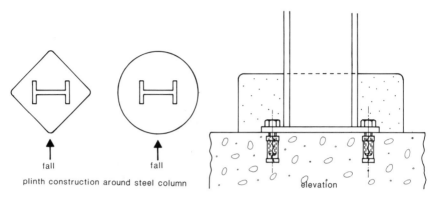

Figure 5.9 Plinth construction around a steel column

then the final use is known—the equipment or pipework passing through can be fitted with a shroud to cover the opening, then so much the better. See Figure 6.2 in Chapter 6.

5.5 Design related to non-cementitious substrates

While by far the highest proportion of substrates comprise concrete in one form or another, steel in particular is a common subfloor generally occurring as decking within buildings or tanker loading platforms and so on and in need of protection.

The main concern when considering superimposing specialist finishes on steel substrates is that undue flexing can lead to fracture of the finish, and therefore unless truly flexible floor finishes are applied, supporting structures should be designed to minimize flexing under optimum working conditions. While the writer would wish to be more specific about the acceptable levels of flexing the reader will realize that the wide variety of flooring systems, monolithic and tile, and in addition the thickness of the finish, will influence tolerance of flexing. Be aware of the need to control it, consult with specialist contractors and discuss in relation to the specification offered what tolerances are required.

It is common practice on steel deck floors to leave plates unjointed over supporting beams. This practice should be discouraged when finishes or toppings are to be applied and plates should be tack welded to each other and to the beam, or countersunk screwed to the beam. Welding the underside of plates to the 'I' beam causes curling of the edges over the beam and formation of a potential flexing point and should be avoided. Figure 5.10 indicates the preferred detail of floor plate fixing.

Timber floors are rarely contemplated and are generally existing substrates. All of the comments applicable to steel decks in relation to flexing apply to timber. In particular board joints must be designed not to move.

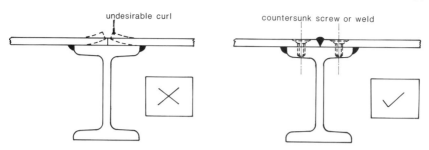

Figure 5.10 Preferred detail of floor plate fixing

5.6 Preparation of new substrates—concrete

If proper consultation has been made with specialist contractors it is quite possible that the cost of preparation to receive topping or finishes can be reduced or eliminated by one of the parties involved in the overall construction. However, various options are available to produce acceptable substrates from those less than desirable, as laid.

5.6.1 *Vacuum shotblasting*

In finishes such as resin toppings where structural integrity is essential with the substrate, vacuum shotblasting is an increasingly popular method of surface preparation. It is, however, generally considered only cost-effective on large open areas due to the considerable size of the equipment involved, and the cost of its operation. Where it is to be used, civil contractors may find themselves held to less rigorous tolerances on the texture and nature of the immediate substrate surface.

5.6.2 *Scabbling*

If design considerations have been applied, it is highly unlikely that new substrates need to be subject to scabbling unless levels are to be corrected (see preparation of existing substrates). Scabbling may be considered as a valid alternative to vacuum shotblasting on small areas if the necessity for preparation arises.

5.6.3 *Acid etching*

The acid etching method always seems to attract controversy. However, the writer has experience of many situations where this has been used to good effect.

Properly used in situations where for instance vacuum shotblasting is not possible and where laitance is present, the use of 5 percent hydrochloric acid with a thorough rinse is far less likely to promote failure than the presence of the laitance itself. Furthermore this method, properly executed, does not saturate the substrate to the extent that the surface finish cannot be applied within a reasonable time as is sometimes believed.

5.7 Existing substrates

In considering application of specialist finishes to existing substrates, limitations are imposed by the design and condition of the substrate in question. Furthermore, many floors considered for refurbishment are contaminated, degraded or have an existing finish which is unsuitable. The degree to which any remedial preparation is required will be related to the specification. For the purpose of this section, all aspects will be covered as a matter of course.

If we deal first with removal of existing finishes, we can then move on to the assessment and preparation of an existing substrate to receive new finishes.

5.7.1 *Removal of existing finishes*

Paints and sealers. Unless paints and sealers are present in only localized areas (in which case they can be treated as contamination) the use of chemical paint strippers, even water miscible types, is hardly a practical proposition. The use of vacuum shotblasting and/or scabbling has in the main proved effective for removal of paints and sealers and these methods are discussed separately. Both methods avoid the danger of residue from chemical cleaning methods affecting subsequent finishes as will occur if they are not thoroughly removed. Furthermore, physical removal, while creating potential dust problems is sometimes preferred to chemicals promoting strong odours which are difficult to isolate.

Thin resin toppings through pavior finishes. There is little alternative in removal of thin resin toppings finishes than to use electric or pneumatic hammers fitted with spade chisels or other suitable tools.

Having removed existing finishes and cleaned down the substrate, a thorough inspection of the floor areas should be made with the following parameters in mind.

(a) The substrate must be sound and free from the effects of chemical attack and contamination, or capable of reinstatement.
(b) Contamination by oils, greases and solvents must be capable of complete removal.
(c) Freedom from the effects of rising damp or hydrostatic pressure should be ascertained.
(d) Arrises at construction joints and expansion joints must be square and sound or made so unless isolated screeds are to be superimposed.
(e) Repaired areas, if present must be to the standard of the surrounding floor area or made so. Core samples are best taken to determine the condition of

repair areas as they are unlikely to be included on original specification or drawings, or to be to the same specification as the main floor.

Existing substrates in poor condition should only be considered as a support for a new screed laid on an isolating layer but in this respect they must still comply with design requirements unless the new screed design ignores any structural benefit from them. Similarly, if the falls or drainage are inadequate a new screed will be necessary.

Having assessed the condition of the substrate, various undesirable elements may have been identified and be in need of correction.

5.7.2 *Moisture*

Moisture is not only a problem in new cementitious screeds. Its presence in existing mature substrates can also lead to failure of finishes by various means such as:

(a) Volatization beneath resin toppings under temperature influences, creating blistering
(b) Prevention of setting or full cure of sensitive materials
(c) Chemical reaction with urethane systems
(d) Influencing of the physical nature of the substrate which can change with varying water content.

In (a) water trapped beneath relatively impervious resin toppings will, under the influence of temperature, even direct sunlight, result in blistering of the topping. Volatilization of the water proceeds at a rate greater than the rate of permeation through the finish. At the same time the toppings may become plastic with the temperature and have a subsequently low distortion pressure. Blisters thus formed can be broken down by traffic at a later stage, particularly when the temperature has fallen.

In (b) furane, polyester and certain epoxies may be retarded as a result of moisture contamination.

Presence of moisture with urethane systems (c) generates carbon dioxide gas and contaminated materials have a porous texture.

By (d) saturated concrete has a greater mass than dry concrete. Finishes applied to wet concrete, even those compatible with moisture, will be subject to stress when and if the substrate eventually dries out. Substrates exhibiting fine surface cracks or early age shrinkage cracks before installation of a finish influence the topping if the water content or temperature varies, promoting alternate expansion and contraction which focuses at weak points on the substrate surface and reflects in the finish.

5.7.3 *Removal of contamination*

Oils and greases. Removal of oils and greases is essential to successful refurbishment of floors. Oils and greases will prevent adherence, and may subsequently volatilize to cause blistering, certainly to resin toppings. If allowed to come into contact with membranes based on bitumen or natural and synthetic rubbers, they will cause irreversible degradation.

Because this form of contamination generally occurs on a local basis, methods of treatment may be attempted which may not be practical on large floor areas. Degreasing solvents may be used, but may also carry the contamination deeper into the substrate, therefore they should only be used in conjunction with high absorbancy powders to draw out the contamination. Degreased areas should then be subject to the preparation used to clean and prepare the general floor area.

Paint. On a local basis paint splashes can be removed with paint strippers and a sharp scraper, or if dust is not a nuisance by abrading with a stone cutting disc.

Chemical contamination. The potential contaminants are so diverse that generalization is essential if it is to be meaningful. Residual chemicals in substrates will interfere with the adhesion of subsequent finishes and may also continue to degrade the substrate by chemical attack long after a new finish has been laid. The basic intention is to remove the degraded substrate to an extent where a surface is exposed which is strong enough and clean enough to support a new finish or be repaired and maintain long-term integrity.

Concrete substrates subjected to chemical attack by acids are broadly affected in two ways.

(a) By reaction with the cement content to form salts having a greater volume than originally as laid. This form of attack will cause floor failure of floor finishes by 'blowing off' the topping. Residual concrete attacked in this way can be described as having a 'mushy' consistency.

(b) By dissolution of the cement content, breaking down the concrete structure leaving a sandy texture or areas of eroded surface with proud aggregate. Occurring beneath finishes, this form of attack undermines the support given and the finishes collapse.

In both instances, the most effective method of removal of degraded material is high pressure water hosing, otherwise known as waterblasting. This method is highly efficient at removing contamination and loose aggregate, but as most of the product removed ends up in the drainage systems, much of the contaminated surface as can be removed manually should be taken off, to reduce the possibility of blocked drains. Vacuum shotblasting is only effective in dry situations which are uncommon in remedial work.

If large areas of the substrate are corroded, core samples should be taken to determine whether a satisfactory substrate remains. If this is the case, or where corrosion has been light, neutralization of the surface should be carried out.

Neutralization is best carried out by use of materials which in themselves offer least danger to subsequent finishes or to personnel, and in this respect the next best material to water alone is a dilute solution (10%) of sodium carbonate (soda ash) brushed into the surface and left wet overnight to neutralize residual acids in the substrate rather than just the surface contamination. The surfaces should then be hosed with clean water and thoroughly brushed. When the surfaces are dry they should then be brushed again to remove sand and other debris.

Whether or not the surface is now suitable to receive a finish will depend on the specification being considered. It should not be assumed that a single treatment of badly contaminated surfaces will suffice unless a qualified person has made the assessment.

Concrete subjected to alkali attack. While alkali attack on concrete is not as pronounced as acid attack, it is nonetheless important to dilute the contaminant if future crystallization is not to occur with subsequent pressures which may rupture light finishes. Without specialist advice the use of dilute acid solutions to counter presence of alkali is not recommended and as for corroded concrete in general, water blasting will generally adequately neutralize the surface to the normal levels of alkalinity that concretes possess.

Whenever utilizing chemicals in treatment of substrates, proper attention must be paid to protective clothing and manufacturers' recommendations.

5.8 Repair of prepared or existing substrates

Following preparation, most existing substrates are in need of repair of some nature.

The most common areas of damage, excluding general areas of corrosion, are arrises of construction joints and movement joints. If it is intended that finishes of any type are to be superimposed directly on to the prepared substrate, then repair must be effected. It is not acceptable to install battens to the joints and allow the finish to fill naturally any broken edges. If this is allowed to happen, the chances are that the finish will at some point be applied over an unsound edge with subsequent failure due to lack of support. All suspect edges must be chipped away to sound material and reinstated utilizing a suitable material compatible with the finish, as diagrammatically indicated in Figure 5.11.

For isolated pitting, repairs should be carried out on a local basis, again taking into account the nature of the finish to be applied.

Where extensive remedial work is required to the surface, or where rescreeding to falls is necessary, a levelling screed is recommended to provide a suitable surface for resin toppings or thinly bedded tiles.

5.8.1 *Suitable materials for repair of existing substrates*

The main criteria for repair are that first the repair material must be capable of bonding to the substrate under the prevailing conditions, and remain bonded

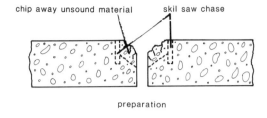

preparation

completed reconstruction of arrises

Figure 5.11 Substrate repair: (a) preparation; (b) completed reconstruction of arrises

during the life of the finish without cracking. Secondly the repair material must be capable of compatibility with the finish. If the repair material is not chemically compatible with the finish then the design of the repair mix must be such that a mechanical key is produced by way of a priming system. Suitable materials for repair of substrates are:

(a) natural or synthetic latex modified cementitious products
(b) cementitious compounds with suitable bonding agents
(c) epoxy resins, whether or not modified with cementitious fillers
(d) polyester resins subject to compatibility
(e) any other product tried or tested by the supplier of the finishing material.

A point worthy of note is that with the latex modified cementitious products, if the repair or screeding mix is not properly ratioed and the mix is too sloppy, trowelling produces a rubbery fatty skin of weak strength and low adherence potential. The preferred method of application is to apply a semi-dry mix to a wet primer and ensure a good compaction of the repair material. Finish with wood float to achieve a fine but fat free surface.

5.9 Existing steel substrates

Existing steel substrates must comply with the design requirements for new steel substrates for a given finishing material.

Corroded steel substrates are generally easier to repair than concrete structures in that the welding in of new plates restores the surface completely without concern for compatibility of materials, subsequent hairline cracking or other concrete repair problems. Without doubt, however, the need to remove all corrosion products is paramount.

6 Design of finishes

In so many specialist flooring applications, it is the thought, knowledge and skill that goes into the design and application of the specialist finish that sets different contractors apart. The contractor is the final link in the chain and it is incumbent upon him to temper the imagination of consultants and architects into practical workable specifications and not to following blindly a specification without those attributes. It is hoped that this section will enlighten those writing specifications and reduce the gap between theory and practice.

Many end users will appreciate that purchasing an epoxy resin topping for instance, or acid resistant tile finish, does not guarantee a floor fit for purpose. Simply to use a chemical resistant product without thought for proper site and substrate preparation, design of movement joints, skirting, membrane terminations (if included), drainage method and the physical and chemical requirements of the floor finish, will lead to dissatisfaction. All of these various aspects of the successful installation of specialist finishes are covered separately. In this section we hope to identify the aspects of design of the finishes themselves which lead to a successful application. No apology is made for repetition of salient points highlighted elsewhere—most need to be assured of notice.

6.1 Product design—screeds

6.1.1 *Polymer modified cementitious screeds*

As these materials are used primarily for non-chemical resistant situations, it might be thought that their design would not be so critical. In the sense of comparison with tile systems for chemical duties, that might be so. However, a customer requiring a polymer finish requires a system just as durable as the customer requiring full chemical protection, and particularly if the finish is required to be hygienic.

Product design is important, and as with any product, the easier it is to apply, the better the resulting finish. The best systems for finishes do not utilize building sands as might be used in the substrate or in simple levelling screeds, but use clean washed and dried silica sands, blended and prepacked, with the necessary ratios and grain sizes of sands, cement and aggregate, carefully controlled. Materials that are required to be applied as thin sections will contain no coarse aggregate, particularly those where feather edge finishing is required. *Note*: Where edges are required to be finished to a feather edge, it is recommended that a 25 mm × 25 mm chase is formed to reinforce the terminating edge.

The mix design is important, but varies with the type of polymer liquid, type of sand and type of aggregate. The starting point for mix design is the polymer: cement ratio, which for styrene butadiene copolymers is between three and five parts cement by weight to one of polymer, and with acrylics between two and five parts.

For basic renders, sand: cement ratios are 3:1 and for aggregated systems 2.5 to 3 parts of a 50/50 sand–aggregate mix with 1 part cement. Water is added only to control consistency and workability up to a maximum 50% water: polymer ratio. When formulating semi-dry systems, enough liquid must be present to enable compaction under the condition of application; it is not adequate to simply close the surface. Semi-dry systems should always be laid on a wet primer, to provide a link between substrate and topping.

Anti-foam materials are normally added to the liquids to minimize air entrainment in the final mix, which is responsible for porosity particularly in machine mixed systems. For other than basic levelling screeds, only clean washed and dried silica sands should be incorporated in specialist finishes. Over-use of fine particles such as pigments should be avoided.

Aggregates commonly used to increase resistance to wear and impact in polymer screeds and toppings are graded flint and or granite up to 3 mm size.

6.1.2 *Resin modified cementitious toppings*

All of the above recommendations for control of quality of sand and aggregates applies to resin modified systems, and more so. Building sands are never used, only clean washed and dried materials. The use of impure materials can have an effect on the resin–catalyst system, which is clearly a critical part of the mix. Proper calculation of the water content is critical as although the resin systems are water tolerant, they do have limits, and there is no doubt that higher water contents produce poorer properties.

The starting point for formulating resin modified topping systems is the resin to hardener or catalyst ratio which will vary with system and manufacturer. Cement to sand ratios will generally be 1:4 as with hydraulic systems. Water will be necessary to hydrate the cement and this is generally predispersed in the resin or hardener system.

6.1.3 *Resin toppings*

In moving on to resin toppings as opposed to resin modified systems, exclusion of all acid soluble matter is essential to enable satisfactory resistance to a wide range of chemicals to be achieved. While the aggregates do not contribute directly to the chemical resistance, they will have some negative effect if they are anything other than pure, and parameters with respect to filler formulation are dealt with in a specific chapter. The resin will be selected primarily as a result of the chemical data provided but there are other requirements to consider such as ambient site conditions, substrate conditions and traffic density, which will have an influence on selection, and the word 'compromise' comes to mind again. The product ratios, particularly with regard to resin and hardener, cannot be readily stated as part of the product design because of the variety of materials and their reactivities, but resin toppings must be formulated with the following in mind:

(a) chemical resistance must be appropriate to the duty
(b) application must be practical over a range of ambient temperatures and in industrial situations
(c) ease of application and ability to close the surface must be built into the formulation just as chemical resistance must be; the best chemical resistant formulation will fail if it cannot be properly applied.
(d) granulometry of the filler must be such that the consolidated topping is as dense as possible and be neither too dry nor wet during the application.

6.2 Installation design—topping

Other than in small patching operations, polymer and resin modified or pure resin toppings are unlikely to be laid at less than 4 mm unless surface self levelling. Generally thicknesses are around 5 to 6 mm for a combination of economy and performance. At these thicknesses the toppings are suitable for light to medium duty traffic which would involve pneumatically tyred fork lift trucks (thoroughfare only: pick up and deposit areas need to be thicker) small hand pallet trucks and so on. Heavy duty aggregated systems are laid 8–20 mm thick (generally resin modified toppings only) and are suitable for areas where mechanical damage is more prevalent.

Resin toppings are essentially designed to follow the substrate to which they are applied and therefore the substrate design and the installation design are closely linked. In new installations designs can be complementary to the finish; in existing installations a compromise must be reached.

Apart from the obvious requirements of adequate thickness, good adhesion and density of finish, the most important design aspects of resin toppings are terminating points and movement joints. Under no circumstances should the design of the finish permit a terminating point at floor level other than at movement joints. Upstands are essential at perimeters, plinths, pipe ducts and service sleeves and Figures 6.1 and 6.2 show various termination details.

Figure 6.1 shows terminating rebates at upstand and movement joint. Upstands are an important feature around service ducts. Figure 6.2 shows finishing detail for screed, membrane and topping or modular finish.

In most instances, termination points are chases which permit localized thickening of the topping finish to strengthen the edge, provide a greater

64 SPECIALIST FLOOR FINISHES FOR CONCRETE

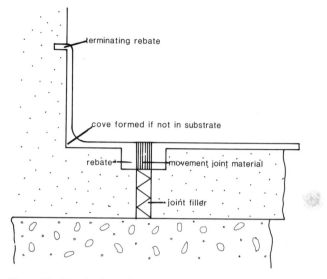

Figure 6.1 Terminating rebates at upstand and movement joints

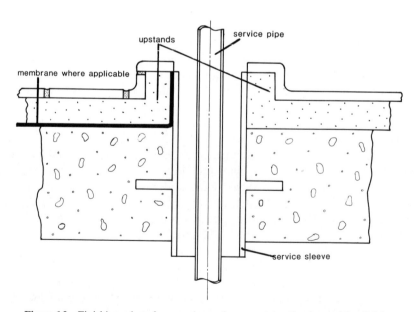

Figure 6.2 Finishing rebate for screed, membrane and topping or modular finish

DESIGN OF FINISHES

bonding area for movement joints where applicable, and to limit possibilities of edges lifting by forming a mechanical anchor.

A separate section has been devoted to movement joint design for all finishes. However, the following rules should be applied.

(a) Always follow movement joints in the screed or substrate with a joint of identical or better movement potential.
(b) Movement joints should be included at all perimeters and fixed points such as plinths, columns and any upstands, plus around any item of plant liable to generate excessive heat by conduction, radiation or by discharge of hot liquor on to the floor.
(c) Movement joints should be constructed in the finish around any plant or equipment generating vibration.

6.3 Product design—modular finishes

In theory, much of the design of modular finishes is done by the tile manufacturer. In practice the success or failure of a tiled floor finish is more dependent on the design and installation of the bedding and jointing system, although that is not to say that floors do not fail through poor tile selection.

There are very few chemical environments which cannot be coped with by a tile system and membrane in a flooring situation, and this property is inferred by the high chemical resistance of the tile body, particularly those tiles classed as vitrified or fully vitrified. The jointing system must be carefully selected as the resins available have widely differing resistances. A broader range of resins are suited for jointing systems than for screeding systems, although in the main epoxy, furane and polyester are most used in specialist flooring.

Tiles may be selected for aesthetic properties by architects and or consultants, with consideration for traffic density at the same time, but the contractor from his experience must decide the method of installation and in what jointing system and method the tiles are laid. The contractor's liability extends to calling into question any specification received which in light of his knowledge and experience is inadequate.

The bedding and jointing system and membrane, where applicable, will be decided on by a whole range of considerations and is far from straightforward. Although experienced specifiers may derive a specification in minutes, the possible permutations actually considered in the mental computer are many hundreds. The various parameters for consideration are summarized as follows for each of the elements of a tile floor finish.

6.3.1 *The tile*

Colour, pattern, module size, thickness, anti-slip or plain, chemical properties, mechanical properties.

6.3.2 The fixing system

Chemical and thermal properties required—which resin?
Fully bed and joint or point only?
Confirm compatibility of resin with bedding or membrane
Floating system or bonded?
Is resin suited to ambient conditions—cold, dry, wet, hot?
New installation or existing?

6.3.3 Movement joints

Requirements—tile and fixing system parameters
Temperature fluctuations or cycling?
Suspended floor or ground floor?
Existing joints in substrate?
Type of traffic—dictates hardness of joint
Do areas require isolating due to temperature or vibration?

6.3.4 Membrane

Chemical and thermal properties required
Installation requirements—new installation or existing in use area?
Sheet or liquid (liquids more sensitive to moisture but more secure), condition of substrate.

6.3.5 The site

Ground floor or suspended?
Inside area or external (subject to climatic variations, rain or direct sunlight)
New or existing installation?
Time of year?

Permutations are almost endless which is why the choice is very specialist in nature. Add to the above the possible variations in application skills and potential changes of use, and it will be understood why problems can and do arise.

The responsibility for design of the tile lies clearly with the manufacturer, and each publishes clear information to enable selection to be made. Specification problems generally arise as a result of underestimating the physical requirements of the floor, or because the use to which the floor is put has been changed or added to. One specific point to remember is that tiles less than 50 mm thick, laid as a floating system, must be bedded to equal that thickness or more. If, in a chemical resistant situation a membrane is required directly beneath the finish, the minimum 50 mm thickness applies unless the tile is elastically bonded. These thicknesses apply irrespective of traffic density.

The design of the fixing system is related to the other elements of the specification selection. Bedding materials must have adequate compressive strength when mature to withstand the applied loads and shock loads transferred through the tile during use and to provide the adhesive properties necessary to provide a bonding bridge between the substrate and tile or in a floating system between itself and the tile.

It is also important that bedding systems are designed to give a necessary degree of support during installation and before setting, which allows the tradesmen to fix the module correctly, and for it not to change its position and require frequent resetting. This in itself can lead to premature failure if adjustment is carried out at the wrong time, as cement based systems are very weak when immature.

Semi-dry screeds are used most frequently in dairies, breweries, bakeries, food processing and most hygiene installations. The tiles are fixed to level integrally with the screed, eliminating the requirement of a separate screed to falls and this method has advantages in low shrinkage properties compared to concretes with normal water contents. As is mentioned several times in sections related to semi-dry screeds, compaction of the screed is just as important as correctly ratioing the sand: cement content and ensuring proper mixing. Badly compacted semi-dry screeds have minimal strength, and crumble in a manner like screeds with a low cement content.

Semi-dry screeds (from data provided by Ceram Research) can take up to one year to reach full compressive strength, with 70% developed after 14 days.

6.3.6 *Tile joints*

The jointing materials used with tiling systems either as an integral bed and joint or as a joint only should not be underestimated in their requirements.

At the risk of stating the obvious, tile joints:

(a) join together individual tile modules
(b) prevent ingress of chemicals, moisture and dirt, and bacteria between tiles
(c) support the edges of the tile module during use
(d) provide a facility for absorbing dimensional tolerances
(e) act as a buffer between tiles, absorbing certain physical stresses.

Jointing mortars need to be formulated in accordance with the method of application in mind, as clearly those for hand pointing or buttering must be stiffer than those required for gun jointing. The more fluid gun jointing systems must not lack thixotropy otherwise slumping will occur creating undercut joints. Subsequent flushing of undercut joints invariably leads to later separation, unless carried out quickly and before the joint becomes dirty, therefore it is important to acieve the correct consistency initially.

Control of mortar plasticity is carried out by variations of filler formulation as discussed in Chapter 3 on fillers and not by simply reducing filler content which can lead to slumping and consequential misalignment.

6.4 Installation design—modular finishes

It is only right that having decided on installation of an expensive floor finish the end user should not be presented with a series of modules laid in such a way as to simply qualify as a floor. Specialist installations must be laid as a system designed and applied with a view to carrying out all the functions of a working platform, a means of discharging aggressive chemicals to drain, a non-harbourer of bacteria and the main means of protection to the supporting substrates and membranes. It is therefore encumbent on the contractor to apply the finishes in a manner conducive to achieving this aim by utilization of proper trade practices and experience in the application, and in guiding of others less aware of the consequences of inadequate installations.

As far as the main floor areas are concerned, the most important design aspect of the installation is the degree of fall and its direction. If they are not already included in the substrate the fall must be such that it brings spillages to the channels or gullies by the most appropriate route to minimize dwell time on the floor, but not in such a manner that they interfere with or restrict movement on the floor because they are too steep. The steepest acceptable fall would be 1:40, the minimum 1:100. In general, the falls in finishes should be designed and constructed as for the substrate as detailed in Chapter 5.

The method of laying 'in bond' can be important in heavy trucking areas and in areas of heavy chemical spillage when the use of industrial paviors are desired. If it can be avoided, trucking and chemical flow should not be along the long joints. In the latter case, erosion of the joint can take place, in the former case a greater degree of edge spalling occurs. Figures 6.3, 6.4 and 6.5 depict the three methods of laying industrial paviors in bond.

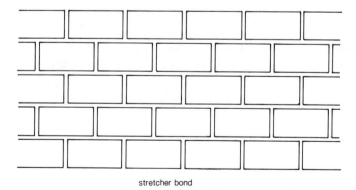

stretcher bond

Figure 6.3 Laying in bond: stretcher

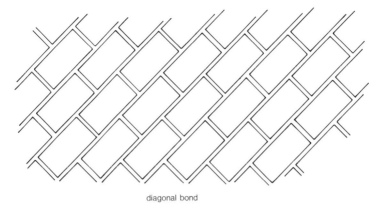

diagonal bond

Figure 6.4 Laying in bond: diagonal

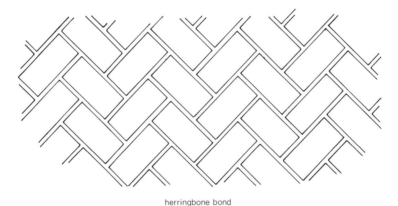

herringbone bond

Figure 6.5 Laying in bond: herringbone

Only herringbone pattern produces a multi-directional floor from both chemical and physical aspects. It is appreciated that to specify and enforce traffic direction on most floor areas is, if not unreasonable, difficult. It can be an important factor in ultimate life of a finish, and as such worthy of note. The direction of laying is well worth considering when very aggressive chemicals are concerned to avoid erosion of joints.

Many more bonds exist for tile finishes.

6.4.1 *Bed thicknesses—joint widths*

Bed thicknesses and joint width must be considered before laying and even at the time of costing. Technically and financially, joints and bed are best kept to minimum thicknesses, but it is very infrequent that these two considerations are other than in the specification or the mind. In practice, the bed must absorb undulations and lack of tolerance in the floor falls—particularly in thin bed systems on screeds to falls—and this can be an extremely expensive procedure.

Joint widths are even more contentious because they are visible. 3 mm joints are commonly referred to as the optimum for joints subject to strong chemical spillage. They are, however, only achievable in fully bed and joint systems and generally at the expense of aesthetics and the line of the 'perps' (end joints, perpendicular on walls) and 5 mm is much more realistic as an average.

Tolerances on module lengths, particularly on thicker modules, must be absorbed by the joint in aesthetic floors if they are to achieve that aim. In these circumstances between 7 mm and 10 mm are not out of order.

When tiles or paviors are set out on semi-dry beds, open jointed for subsequent grouting, then the joint width must be adequate to permit the grouting material to flow to full depth without effort thus avoiding trapping of air pockets. Again 7–10mm is the desirable minimum.

Particular attention should be paid to the layout of the floor in relation to movement joints. Positioning should be determined in advance, where possible, coinciding with high points and avoiding intersecting channels.

An aspect which is appropriate to installation design of modular systems where full chemical resistance is required is the stopper course. When pavior finishes are laid over a membrane to falls, particularly those membranes such as polyethylene or polypropylene which have low adhesive potential, there is a tendency for the finish to slide down the falls as a result of temperature cycling or trafficking. In floor systems with pavior channels and rebates, the floor finish movement results in collapse of the sides. Stopper courses are thicker tiles laid into a rebate in the substrate (but finishing flush with the floor finish) which supports the floor and takes pressures from the channel sides as indicated in Figure 6.6.

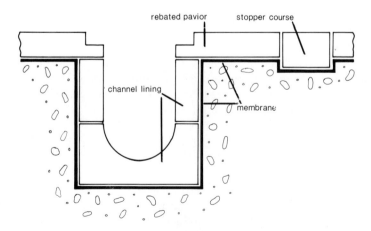

Figure 6.6

DESIGN OF FINISHES 71

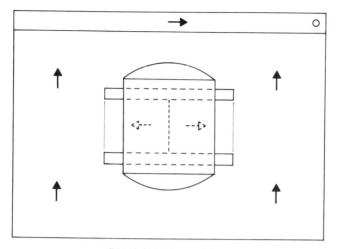

floor pitched between plinths

Figure 6.7 Floor pitch between plinths

6.4.2 *Columns and plinths*

Columns and plinths are generally considered of nuisance value to tiling contractors because of the interruption to setting out they cause. Plinths are the main method of elevating plant and equipment from general floor areas out of the way of corrosive spillages and facilitating cleaning.

The alternative in less corrosive environments, namely stainless steel legs with their point loading, can be just as detrimental to the floor finish after completion as plinths are inconvenient to floor layouts. The method of tiling around plinths that is most effective is that recommended for skirtings, that is to say coved skirting tiles. Apart from producing an easily cleaned and/or neutralized corner, the use of a cove brings the joint, normally an expansion joint, away from the corner and into a position where it can be readily treated.

The height of tiling to plinths or the need to cover completely the exposed surfaces is optional and dependent on the chemical environment and the use to which the plinth is to be put. If a membrane is incorporated, the top of the skirting tile or upper limit of the tiling should seal the terminating rebate where applicable. Where a series of plinths or double plinths such as tank supports are constructed at 90° to the direction of fall, then the area between them should be pitched to throw out liquor into the general falls as in Figure 6.7.

Plinths designed to receive equipment in strongly corrosive areas or in hygiene situations will need fully protecting and those which require holding-down bolts will require special attention if sealing is to be adequate. Plinths to be used for transfer pumps or equipment prone to leakage must be fully protected. Failure to protect plinths can clearly lead to breakdown of the finish surrounding the plinth. Figure 6.8(a), (b) and (c) shows the essential

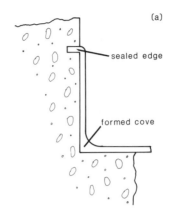

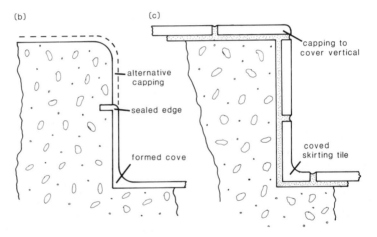

Figure 6.8 Rebates essential for termination of resin toppings on vertical surfaces or plinths

rebates for termination of resin toppings on vertical surfaces or plinths. On tiled finishes capping tiles must always cover the vertical tiles if proper protection is to be afforded.

6.4.3 Channels

Channels are the most popular form of drainage because they enable considerable volumes of liquor to be collected from a large floor area in a short period. Because they are the focal point of all spillages they require special attention in their construction, particularly on suspended floors where they are often the point of leakage. Membranes must be totally sealed because they are more likely to be subject to a head of liquor condition in a channel than

DESIGN OF FINISHES 73

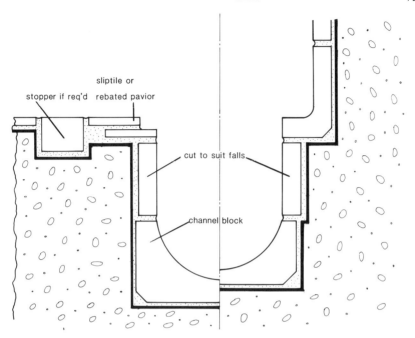

Figure 6.9 Channel section

anywhere else. (See membranes (chapter 4) and installation of finishes, section 9.2).

The channel must be laid to falls in the same way as the floor areas. Channel outlet tiles and blocks are available, all of which have lip or spigot section on the underside to seal into the substrate outlet pipe. Methods of lining channels vary from use of standard bricks or tiles to special shapes. All must be fitted with extreme care with particular attention paid to filling of the joints.

For the most part, open channels are undesirable in most industrial installations and therefore provision must be made for installation of a protective cover or grating. The most appropriate way to achieve this facility is to provide a rebate at the top edge of the channel into which the grating can sit. Thought must be given as to the way this can be achieved, as badly constructed rebates can lead to failure due to the volume of liquor passing over the area. In the same way as with resin toppings, corrosion levels permitting, steel or stainless steel angles can be bedded into the substrate which will accept heavy steel gratings, and these were detailed in Chapter 5.

Rebates can be formed by slip tiles or by rebated tiles or paviors and in these instances the end user must be made fully aware of the need to be very careful as to the method of removal and refitting of grating covers. GRP or polypropylene covers for tiled rebates are less likely to cause damage and are finding widespread use in all industries.

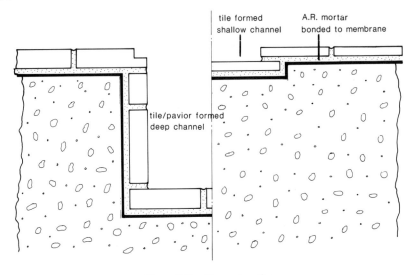

Figure 6.10 Channel section

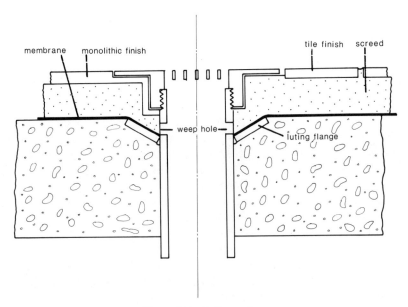

Figure 6.11 Gulley outlets

DESIGN OF FINISHES 75

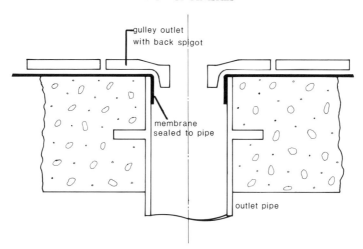

Figure 6.12 Gulley outlet with back spigot

Typical channel sections are indicated below in Figures 6.9 and 6.10. Vertical linings must be bonded to the screed or elastically bonded to the membrane. If loose membranes are used, the lining to the wall should not be less than 50 mm.

6.4.4 *Gulley outlets*

Many flooring situations require gulley outlets rather than channels and in terms of installation detail they require as much attention as channels as they

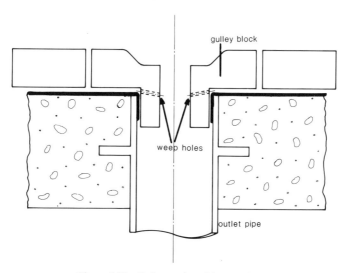

Figure 6.13 Gulley outlet with weep holes

are a common source of leakage, generally associated with the fitting of the membrane.

Where cast steel outlet spigots are used they generally are fitted with a luting flange into which the membrane can be terminated and which contains weep holes to drain liquor from the surface of the membrane layer (Figure 6.11). It is important that the gulley units fitted have a back spigot which penetrates into the outlet pipe. This prevents or reduces the possibility of liquors creeping back under the finish (Figures 6.12 and 6.13).

7 Movement joints

Movement joints are an essential aspect of any floor finish and not only where elevated temperatures are experienced. The range of ambient temperature will influence the frequency of the joints, but all substrates and finishes not elastic in nature will require a joint to absorb:

- Slow permanent growth due to moisture absorption or chemical reaction.
- Slow, gradual temperature variations as occurring in seasonal changes.
- Slow, gradual substrate variations due to change in humidity levels as a result of change of use of building or seasonal changes. This is particularly relevant to areas of high groundwater content.
- Slow to medium rate of contraction due to shrinkage at construction joints in new substrates.
- Slow to medium rate of expansion due to ambient temperature changes between construction and use.
- Fast rates of expansion due to hot water hosing or local high temperature spillages.
- Local vibration of any source.

It is very important to recognize the following:

(a) For a given change in temperature a screed or topping material will expand by a given amount; however, the rate of that change will influence the choice of type of movement joint material and not simply the amount of expansion alone.
(b) The nature of the traffic will govern the type of movement joint material, i.e. heavy traffic areas will require a movement joint which will support the edge of the tile or resin topping, in pedestrian areas softer, more elastic movement joint materials can be used.

It stands to reason that the softer, higher plasticity joints will react more quickly to such temperature changes as occur in areas of high temperature spillage, but possibly contribute to failure of tile edges under conditions of heavy traffic or hard wheeled traffic because of lack of support at tile or topping edges. Conversely, joints that because of their firmness or lack of plasticity will support edges and heavy traffic may fail if the rate of change is to quick, even for degrees of movement within their acceptable limits.

The choice of movement joint materials is therefore not a question only of the degree of movement, but of rate of movement and nature of traffic. All of these factors on a particular floor area will also govern the frequency of the joint. For instance, a floor area subject to heavy traffic and high temperature fluctuations will require firm joints at a greater frequency than a similar floor without the traffic situation.

Note. The use of metal supporting strips for edges of movement joints in heavy traffic areas can only be used where the chemical environment permits. The use of edge supporting strips in combination with high plasticity materials can reduce frequency requirements in traffic areas. Where steel or stainless steel is not suitable for the chemical conditions it may be necessary to use firm support joints more frequently.

The construction of movement joints to a proper specification and in a correct manner is all important. Movement joints have a history of failure in a wide range of floor finishes and are therefore looked on as weaknesses—but they need not be so.

Typical problems associated with movement joints are as follows:

(a) too few movement joints in a given area places excessive stress on those joints present and on the floor finish leading to breakdown of both
(b) poorly constructed movement joints can be completely ineffective if bridged with similar effects to (a)
(c) joints in bonded finishes not properly aligned will create stress at points away from the joint and propagate cracks in the finish
(d) failure to construct joints around local areas of high temperature and/or high temperature spillage such as ovens, pasteurizers, presses and so on will lead to floor failure
(e) incorrect choice of jointing medium will precipitate failure
(f) inadequate joint preparation for bonding will lead to failure under extension by separation and under compression by lifting
(g) failure to ensure sufficient depth will lead to early joint failure, depth to width ratio should be at least 2:1.

7.1 Position of movement joints

Movement joints should be constructed to the perimeters of areas parallel to and following faithfully the skirting and around any intrusions on to the floor area. Plinths, columns, fixed items of plant and machinery should be similarly isolated from the main floor area. The position of the installation of perimeter joints need not be directly against the wall as it may be difficult to install. Provided the joint is installed parallel to the wall it can safely be up to 300 mm away. If the substrate joint beneath is abutting the wall a slip layer will be necessary in this instance between the two joints in the horizontal plane as in Figure 7.1.

The frequency of intermediate joints (those bisecting a floor area) will depend on the factors previously mentioned of temperature, rate of tempera-

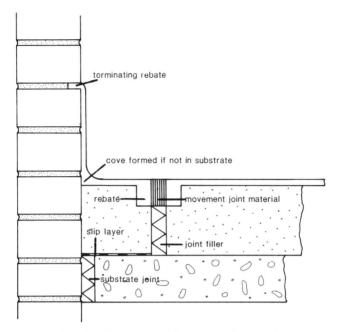

Figure 7.1 Slip plane between two joints where substrate joint abuts wall

ture change, type of traffic and so on, but as a general guide 6 m bays in each direction are appropriate for most installations. On suspended floors 4.5 m bays will accommodate the greater flexing and in this situation the joints should be positioned over supporting walls or beams. CP 202 suggests 9 m bays for tiled floors constructed in semi-dry materials, due to reduced expansion or contraction of these materials.

Without exception, existing intermediate joints in substrates must be repeated in the finish on the same vertical plane, and to an equivalent specification.

With the foregoing excepted, the position of movement joints should be such that they are at high points in the floor. This avoids the dwelling of aggressive liquids on the movement joints, which can generally be of lower chemical resistance than the tiling joints. For this reason, in new installations it is important to take account of the final position of the joints when designing substrates. While on large floor areas it would be impossible to avoid joints in undesirable situations of some degree, but awareness of the requirements can mitigate the problems.

7.2 Construction of movement joints

Movement joints may consist of full depth joints, but for economical reasons are generally constructed as a composite of a cheap back-up material and a

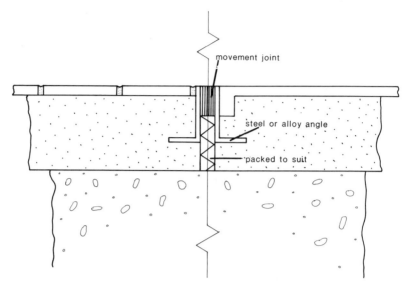

Figure 7.2 Joint details for modular or monolithic finishes

topping of the selected jointing medium. Steel or alloy angles may be incorporated where chemically practical, to reinforce edges of movement joints across trucking lanes and door thresholds and so on. The joint must, as a composite construction, extend the full depth of finish and screed, as applicable, through to the separating layer and/or joint in the substrate. Bridging of the joint by mortar or builders' spoil must at all costs be avoided. Typical joint details for modular or monolithic finishes are shown below in Figures 7.2 and 7.3.

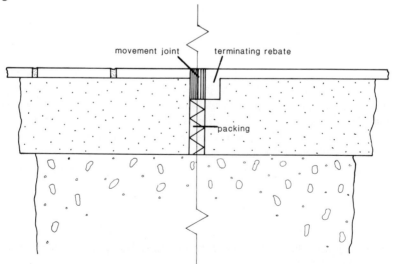

Figure 7.3 Alternative joint details for modular or monolithic finishes

Suitable back-up materials or joint fillers are latex foam, polystyrene, polyurethane foam, fibreboard and bitumen impregnated cork. Where bitumenized compounds or polystyrene are used for back-up, they should be checked for compatibility with the sealant. Bitumens can bleed into jointing materials and polystyrene will melt if in contact with solvent-based primer systems as are sometimes used.

7.3 Common materials used for movement joints

7.3.1 Polysulphides

Polysulphides are most used where the degree of movement expected is considerable. Advantages are in high elongation and quick reaction times combined with good performance at sub-zero temperatures. Chemical resistance and resistance to petroleum is good, but as it is soft, (c. 30° Shore A) support of finishes in heavy trucking areas is limited. Priming of joint sides is required for best results.

7.3.2 Modified epoxies/polysulphides

On the whole, modified epoxies–polysulphides belong to the firm supporting class, lacking in elasticity and speed of reaction, but with a capacity to support edges not possessed by the polysulphides alone.

Hardness, as might be expected, is around 80° Shore A. Chemical resistance is good, and these materials are often used as intermediate joints which are more prone to trafficking. Priming is essential as high stresses can be built up from the lack of elasticity.

7.3.3 Polyurethanes

Polyurethanes have very similar properties to modified epoxies–polysulphides, with perhaps better elongation, but with a distinct problem if moisture is present in the joint. Urethanes react vigorously with water during cure producing a foaming effect resulting in swelling of the joint. Subsequent trimming exposes closed cell pores which are unsightly. Used properly under the correct conditions, polyurethanes offer a good combination of elasticity and support. Hardness is in the range 60–70° Shore A. Priming is essential to seal the joint and provide adequate adhesion.

7.3.4 Silicone rubbers

Silicone rubbers are normally one part, air cured systems with good adhesive properties and good elasticity. They are the products commonly used as household sealants and are generally considered far too expensive for industrial use. Chemical resistance is very good but, like other materials with good elasticity, support is lacking in trucking areas.

7.3.5 Bituminous mastics

Probably the original movement joint material, bituminous mastics are hot poured, and have a tendency to track out of joints and mark light-coloured surfaces. As might be expected, resistance to oils and solvents is limited. Bituminous mastics have high plastic deformation and offer little support, but are widely used as cheap jointing material in substrates. Those substrates which have been in service for any length of time with bitumen movement joints tend to have badly spalled edges.

7.4 Installation of movement joints

The positions of movement joints are predetermined either by design or by the necessity of coinciding with other existing joints or high-risk areas. It is common practice to construct stringer courses or ribs against movement joints in tile finishes, but in resin toppings, it will be formed in the sub-screed with a chase to terminate the topping. In either case, the movement joint backing material is placed against a terminating batten or lath and the screed or tile bedded firmly against it. When the terminating batten can be removed, the back-up material is fixed in place by completing the finish to other side.

In modular installations, the use of supporting timbers when constructing expansion points, ribs or at terminating points, are an essential element in ensuring satisfactory compaction. Hand compaction between wood floats will never produce an adequate compaction on a consistent basis, as without support when the tiles are re-tapped to level, the bedding collapses, losing density. While appreciating the difficulty of utilizing timbers on floors to falls, the improvement in performance is worth the effort, and conversely problems are courted without them.

When tiling and jointing are complete or the resin topping has been laid, the movement joint material can be applied, but not before proper preparation of the joint has taken place. Without doubt, the joint area will have become filled with builder's spoil and cement from the jointing procedures, and if this is not properly removed the joint will be bridged and will not function. Rake out any tramp material, ensuring that any cement adhering to the side of the joint is wire-brushed off. Using a caulking tool or suitable piece of wood, hammer down the back up material until the correct depth for the jointing medium is achieved and finally brush the joint clean.

Certain jointing systems require the use of a primer, and this should now be applied in accordance with the manufacturer's instructions.

Immediately before application of the jointing material the back-up material should be inspected for voids which could result in fluid jointing materials 'sinking' in localized areas. Check that the joint has not become wet or contaminated before application.

For the purpose of aesthetics, it is normal practice to apply masking tape to each side of the joint to allow a clean finish to be created. With tiled systems, care should be taken to ensure that the masking tape does not bridge joints, permitting the jointing meterial to flow underneath the tape at this point. Remove the masking tape as initial setting takes place, for if left too long it will be difficult to remove.

Application of horizontal jointing materials is by pouring after mixing the components with a slow speed drill with paddle attachment. Those materials which generate gases on curing may require a standing period before pouring.

Joints when installed, unless otherwise stated in manufacturer's litcrature, should be finished level with the surface of the finish. Undercut joints will need topping up, with the danger of subsequent separation or lamination where thin sections are concerned. Overfilling will require trimming of the surface which can be unsightly.

7.5 Movement joint calculation parameters

There are few finishes, tile or monolithic, which do not have as part of their technical specification a statement of coefficient of thermal expansion. It generally comprises a number to a negative power which enables comparison with other materials (until somebody uses a different power). These figures are generally quoted for the material or materials in unrestrained expansion which in the case of finishes hardly applies. Furthermore, as previously indicated, even materials having identical expansion coefficients have different amounts of expansion at different temperatures, and at different depths. As would be the case on a floor finish subject to high temperature spillages, temperature gradients as high as 50° C could be set up between the upper and lower surface.

Apart from the problem, therefore, of understanding the methods of calculation and applying them correctly to a defined situation, the layman would find great difficulty in confidently arriving at a solution to the correct decimal place. In order to simplify the calculation, the following typical parameters are offered for a range of materials and composites, as expected approximate movement in millimetres, per 6 m length per 10° C anticipated change in surface temperature. The figures are mean, for the purpose of joint width calculations and represent neither maximum nor minimum movement. The maximum compression or extension values for the specific jointing material should also form part of the calculation of joint width.

Concrete screeds	0.6
Polymer screeds, typical	0.5
Epoxy resin toppings	1.8
Ceramic tiles	0.25
Ceramic tiles on semi-dry bed	0.4
Industrial paviors floating	0.4

8 Site requirements

8.1 Weather protection

Before application of specialist floor finishes can begin, it is necessary to bring the working environments concerned to a standard wherein they can be effectively laid. Very few of the products discussed in this publication, if any at all, can be laid with any continuity or certainty if directly exposed to varying climatic conditions.

Specialist contractors should not be expected to work under the same exposed conditions as say the civil construction contractors providing the substrates or building, but they are, and the end result is not at all satisfactory.

Weather protection for both stored materials and areas of work is essential, irrespective of the ambient conditions. On large open floor areas, this protection can be a considerable item of cost and it is common practice for all participants in a contract to exclude weather protection from the cost and deem it necessary in additional clauses. The net consequence is that nobody allows for weather protection and when the cost is fully appreciated, it ends up as an unsatisfactory compromise between price and partial shelter. Subsequent problems with finishes are often blamed on poor weather protection.

The need for weather protection is equally important in the summer months or hot climates. Direct sunlight will promote blistering in resin toppings, and overdrying of cementitious screeds with subsequent surface crazing. Tiled finishes may expand while the bedding material is fluid and then contract on cooling, placing joints in tensile stress and the bed–tile interface in shear.

Covered buildings solve the weather protection problems except that during winter months ancillary heating will be needed to maintain ambient temperatures to the level required by the particular product. The cost of this can be considerable and must not be forgotten.

8.2 Summary of site requirements

A summary of the site requirements for satisfactory application of specialist resin toppings and tiling would be as follows:

Ambient conditions. Extremes of temperature must be avoided although the degree of control will vary from specification to specification. Certainly, epoxy resins, polyester and polyurethane require in excess of 50° F, 10° C, below which the trowelling and flow properties may be affected, not to mention setting.

Contamination. Water, dust, oil and chemicals will adversely affect resin toppings or tiling joints before full cure being achieved and, for that reason, areas of work should be protected from outside influences.

Freedom from other trades. Particularly on construction sites it is appreciated that maintaining freedom from other trades is not a simple matter. Flooring contractors generally need large areas in which to work for setting out and local storage.

They are also working on areas difficult to avoid, floors, and it must be recognized that severe damage can occur to floor surfaces before maturity, and unjointed tile finishes can be permanently damaged without the problems manifesting themselves until the floors are in use. It is therefore important to give the specialist contractor freedom to work without interference until the floor is released as complete by them.

Storage. Inflammable materials must be stored in accordance with the Health and Safety at Work Act. All other materials must be stored in cool, dry conditions and away from extremes such as frost and direct sunlight.

Protection of finishes. It is very important to protect specialist finishes of all types from subsequent operations before actual use. Effective protection from paint, lagging and so on can be achieved by simple polythene sheeting. However, pipe fitting, welding, bricklaying, roofing, fork lift trucking, craning and other operations involving heavy loads or potential impacts will require that a more substantial protection is utilized, such as hardboard or plywood with joints taped.

9 Installation of specialist floorings

Installation of all materials discussed here requires the use of experienced and skilled tradesmen and for that there is no substitute. Every manufacturer has specific installation details for their own membranes and finishes and, as it is not possible to cover each of them in depth, the following application details are therefore a generalization of trade practices, designed to enlighten the uninitiated and to identify technically sound procedures which apply to the general range of products and their installation, to those whose experience of specialist flooring is not complete.

In this section it is assumed that the substrate, whether slab or screed, has been laid to the standards appropriate to the finish and in accordance with the details given under design of substrates (Chapter 5), and therefore, where applicable to the specification, the membrane will be the first item to be installed.

9.1 Application of membranes

Before laying any form of membrane particularly to concrete structures, the surfaces must be thoroughly brushed or preferably vacuum cleaned. Grit trapped beneath the membrane can cause local thinning and possible premature failure under the applied load of the floor finish. Brush again before application of each panel or areas of adhesive and keep a constant watch for foreign matter being introduced by working operations.

9.1.1 Sheet membranes

Irrespective of whether the membrane is laid on a flat slab or to falls, the position of the drainage outlets will govern the starting point for the installation of the membrane. Membranes are always installed working away from the point of drainage, so that the overlap is not against the direction of flow (Figure 9.1).

Where channels are involved, the start of membrane laying will be at the invert of the channel. The width of the sheet should be continuous around the girth and in as long a length as possible. If the girth of the channel is greater than the sheet width, the invert of the channel should be lined with the sheet centrally positioned so that the joint for the next sheet(s) is not on the invert but as far up the side as practical. Joints should not be made in the bottom of the channel, unless the sheets are not long enough to place in one piece, as may

INSTALLATION OF SPECIALIST FLOORINGS   87

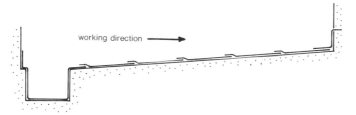

Figure 9.1 Installation of sheet membrane, working away from point of drainage

occur in a long length of channel. Where joints are unavoidable they should be given particularly careful attention.

Outlets are obviously more difficult. However, sealing can be made much more effective if a sleeve of stainless steel or plastic with a thin flange of maximum 2 mm thick is inserted into the gulley and sandwiched between two layers of membrane. The top layer of this sandwich should always be the main membrane layer, thereby ensuring the joint is not against the liquor flow (Figure 9.2). Firstly, a ring of membrane is cut three times the diameter of the outlet and positioned and adhered centrally. The outlet hole is then cut out. The flanged sleeve is placed into the hole or outlet pipe and the flange pressed into a primary ring of adhesive mastic or membrane, as flat as possible.

Laying of the main membrane sheeting follows with the sheet again laid centrally over the outlet, using a suitable bonding medium if the membrane is not self-adhesive. The slight lip formed by the insert flange is insignificant compared to the consequence of a badly fitted and leaking outlet. The outlet hole can then again be trimmed out. If the nature of the membrane permits turning it into the outlet, this can be done, although with a sleeve inserted it is not essential.

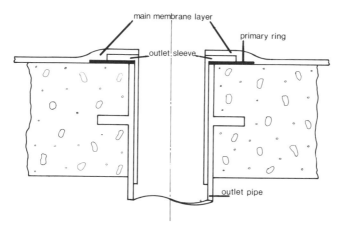

Figure 9.2 Flange inserted between membrane layers to aid sealing

G

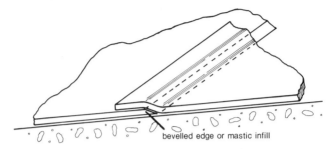

Figure 9.3 Bevelling of edge of sheet

Installation proceeds in each direction, paying particular attention to the cleanliness of the joint before making the final seal. End of panel joints must be staggered by at least 150 mm and preferably by 50% of the panel length. While all joints should be closely inspected, the most common source of problem and leakage is at the end of panel 'T' joints and also where adjacent panels overlap 'T' joints at which point capillaries can be quite easily formed. They are best avoided by bevelling the edge of the sheet to be overlapped or by application of an approved mastic before the joint is made which is then squeezed out to fill any possible capillaries (Figure 9.3).

At wall perimeters and all upstands, the membrane should be turned up without jointing if at all possible. Cutting is obviously necessary at internal and external corners and suitable cover patches must be placed over the joints so formed, overlapping by a minimum 75 mm the extremity of the cut.

Termination of the membrane at walls, perimeters or fixed points, must be into a rebate which is subsequently sealed, as in Figures 9.4 and 9.5.

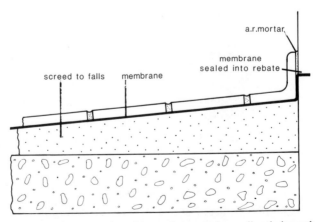

Figure 9.4 Membrane terminated into rebate, membrane directly beneath tiles

INSTALLATION OF SPECIALIST FLOORINGS 89

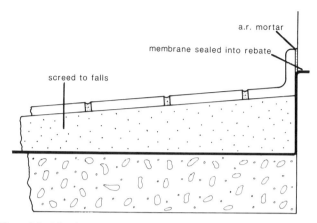

Figure 9.5 Membrane terminated into rebate, membrane beneath screed

9.1.2 *Application of liquid membranes*

Before application of the membrane or the primer, the surfaces must be throughly brushed or preferably vacuum cleaned. Particles of foreign matter may become entrapped in the membrane during application, potentially creating leaks. At best, inclusions will lead to localized thinning of the liquid coat. Attention to cleanliness should be continuous throughout application as dust and grit are continually being reintroduced during work periods.

Liquid membranes applied at 3 mm thickness or less should be applied in two coats to lessen the chance of an inclusion large enough to create loss of integrity. Each coat should be inspected for blemishes and repaired. With liquid membranes, starting points are not as important as with sheet membranes as they are laid essentially jointless. It is, however, good working policy to lay all membranes outward from the drainage point. If then a joint is necessary at some point in the laying exercise then the direction is always correct.

9.1.3 *General*

From a continuity point of view, it would always be appropriate to complete membranes before starting to lay a screed or finish. The whole area can be maintained under proper conditions and the cleaner the area is kept, the more consistent the jointing procedures are.

There are circumstances, however, particularly when laying tile finishes over a large membrane area, when completing the membrane may be counterproductive if areas of it are exposed to trafficking in subsequent laying operations. In this event, sections of the membrane should be laid in limited but convenient areas and the finish laid to protect it before proceeding with the next application. If this procedure is to be adopted, the free edges of the membrane must be protected from other operations by polythene sheeting to keep it clean and by hardboard if there is danger of mechanical damage. Under no circumstances should free edges be subject to contamination or trafficking.

Membranes having high plastic deformation such as polyisobutylene are very sensitive even to kneeling pressures and footmarks, therefore soft sponge kneeling mats should be provided for slow-moving practices such as tiling, involving high point loads.

The final operation before starting to lay a finish over a membrane should be a brush-off of the area and a thorough visual inspection. Repairs are simple at this stage and virtually impossible later when finishes have been applied.

9.2 Application of finishes

9.2.1 *Polymer modified cementitious finishes*

While mixing ratios and proportions may vary from product to product, laying procedures are generally consistent. For other than small quantities, machine mixing is the preferred method, open pan forced action horizontal mixers being suitable, as are trough mixers. Free fall mixers are not recommended as they promote balling of cement with subsequent loss of distribution and weakening of the mix. When mixing mechanically, the use of anti-foaming agents is normally recommended to reduce air entrainment—specific products are generally referred to by manufacturers. The dry ingredients are mixed in the mixer first, then the polymer including anti-foam are added and mixed for 2 to 3 minutes. Overmixing may result in air entrainment, and the mixer blade should be moved regularly to ensure all ingredients are thoroughly mixed to minimize dead zones, and to speed mixing.

It is important to avoid over-wet mixes, which are difficult to level and produce a weak fatty surface. Screeding mixes are generally friable when loose, but will compact and cohere when squeezed in the hand without breaking up again when the pressure is released.

Priming. Dry substrates should be water saturated before priming and all surface water brushed off. Immediately before laying, the substrate should be primed with a slurry of cement and polymer base. Porous substrates, evidenced by over-quick drying, should be coated twice, preferably allowing overnight drying between priming coats. Priming not only provides a sound link between the finish and the substrate but reduces the suction, which will interfere with adhesive properties. Priming must also be carried out to the edges of previously laid bays unless these form the edge of a movement joint.

Application. Laying of the mixed polymer proceeds over the wet primer, the thickness of application being controlled by use of gauging battens between which the material is poured, rough trowelled and then levelled with a board between the battens. Wood float or steel floats are finally used to provide the surface required.

Bay sizes should be predetermined, based on the substrate bay sizes. A 'wet edge' should be kept until these areas are reached and finishing should always be to a straight edge or gauging batten. Laying of alternate bays will reduce the effect of shrinkage on the construction joint, even though shrinkage is nominal and less than in conventional cement mortars. In specialist finishing, any procedure which is likely to benefit the final finish should be adopted. If hairline cracking at or over substrate/finish bay joints is likely to be a problem in subsequent finishes a thin concrete saw cut is made along the joint line later and filled with a firm supporting expansion joint material.

Failure to cut inducement joints may lead to irregular cracking which is difficult to correct at a later stage because the crack cannot be followed with a saw cut. Movement joints in the substrate must be followed without exception, and the width of the joint in the topping should be no less than the joint width in the substrate.

Curing and drying. As with conventional concrete screeds, drying rates should be controlled, particularly in summer months and in situations of controlled low humidity. When the polymer topping has set sufficiently to take weight without marking, wet hessian should be laid on the surface and kept moist for 24 hours and then replaced by polythene sheeting for a further 48 hours.

As a general rule, foot traffic is acceptable within 24 hours and light traffic after 72 hours. Heavy traffic should not be allowed until 10 to 14 days after laying.

9.2.2 *Resin toppings*

Mixing ratios will vary with the type of resin/hardener and with the filler and aggregate formulations. Free fall concrete mixers are not suitable for mixing resin toppings. The best types are compulsory mixers such as horizontal pan mixers with internal fixed scraper, or with double planetary rotating blades. Batch sizes depend on the nature of the area to be laid; large open areas will accommodate large batches and minimize the effect of batch to batch variations. Constricted areas with channels, plinths or upstands will need smaller batches in view of the greater effort required in laying.

Most resin toppings comprise resin and hardener and the correct proportions should always be added to the mixer first and mixed for 1–2 minutes before adding a small quantity of filler which aids blending. The full quantity of filler or aggregate should then be added gradually and mixed for 2–3 minutes, depending on the setting time of the particular product. Thorough wetting of the fillers by the binder must be achieved, and no dry material should remain around the sides of the mixer. If dry material enters the topping, porous areas will be created. The mixer blade should be moved regularly to ensure complete mixing and minimize dead zones.

Priming. Thoroughly brush or vacuum areas before priming. Priming of the screed or base is always recommended even with low viscosity surface self-levelling systems. Priming not only provides the link with the base, but also seals the surface, precluding air inclusions in surface self-levellers, and identifies porous areas of the base by the nature and rate of its drying, allowing re-priming where necessary.

With surface self-levelling systems, the primer is allowed to become dry before laying proceeds as this prevents pick-up of the primer in epoxy systems, and solvent trapping in polyurethanes which use solvent based primers. In resin topping systems, laying proceeds on to the wet primer because the toppings themselves are not sufficiently resin rich to adhere, and require the wet primer as a bridge. The tacky nature of the primer aids consolidation of the topping by preventing product slipping about under trowelling pressures.

Application. (i) Surface self-levelling systems: these compositions are applied by serrated trowel, squeegee or comb. The pitch and size of serrations will govern the final thickness of the coating. The mix is designed to flow to give a smooth surface; however, it must have some thixotropy if it is not to flow down falls. Immediately after trowelling to thickness, a spiked plastic roller is traversed over the surface, releasing trapped air and promoting the surface flow and release of surface tension necessary to achieve a smooth surface. It is essential to avoid spiking on partially set areas otherwise permanent marking will occur. Bays should always be finished to a straight edge, if possible coinciding with a substrate joint. Keeping a wet edge until these points are reached may well require some flexibility of working periods.

Construction joints in the concrete substrate should not be ignored and should be cut out as inducement joints after the finish is laid. Filling of the joint should be with a firm supporting movement joint material such as flexible epoxy or polyurethane. Movement joints must be included in finishes where they are present in the substrate and must be designed to have the same or better movement potential.

Cleanliness during laying of surface self-levelling systems is very important because low viscosity systems are laid thinly. Dirt inclusions will represent high proportions of the finish thickness and surface imperfections are easily seen. When solvents are used for cleaning spiked rollers, before re-using ensure all solvent is removed otherwise marking of the surface will result in the form of spots of a lighter colour. They are, however, only detrimental to the aesthetics of the finish.

Temperatures under which surface self-levelling materials are laid are vital to the effectiveness of the finish. Flow properties are related to temperature of application and in cold conditions pre-warming of the materials is recommended if the environments cannot be controlled.

(ii) Trowelled resin toppings: Unlike surface self levelling systems, these compositions require compaction and closing of the surface by trowelling, and in most instances surface sealing.

Application to wet primer proceeds by trowelling or floating between gauging battens, placing with a sweeping motion and consolidating at the same time. Normally two or three batches are laid and then finished with a steel float to a consistent density, and the mixed material should be adequately plastic to permit this. Application properties will vary with ambient temperature and control of temperature is important as indicated in site requirements. The lower the temperature, the more difficult it becomes to merge batch with batch, with the result that overtrowelling occurs at these points, showing as definitive lines of a wetter finish giving a patchy appearance to the finished topping. The need for a surface sealer will vary with system. However, most trowelled systems benefit from a resin sealer, to improve impermeability and reduce tendency to entrain dirt particles.

Bays should always be finished to a straight edge, where possible coinciding with a substrate joint. Construction joints in the concrete substrate should not be ignored and should be cut out as inducement joints after the finish is laid. Filling of the joint should be with a firm supporting movement joint material such as flexible epoxy or polyurethane. Movement joints must be included in finishes where they are present in the substrate and must be designed to have the same or better movement potential.

(iii) Epoxy terrazzo: Epoxy terrazzo systems are a decorative form of resin topping applied in the same manner as conventional resin toppings up to the trowelling stage, at which point an additional process of grinding takes place. The grinding process exposes and polishes selected aggregates to give a decorative appearance.

Grinding must be carried out within 48 hours, preferably around 24 hours, otherwise the curing of the resin may have reached a stage of hardness which makes grinding difficult. The initial grinding stones are coarse textured carborundum, which, in addition to cutting the aggregates, exposes air pockets from the mixing and laying procedures. When coarse grinding is complete a fine resin paste is squeegeed over the surface to close the pores and fine carborundum stones are then used to produce a fine finish. Sealing of the surface with clear resin sealers improves resistance to dirt and fills small imperfections from the grinding processes.

9.2.3 *Application of modular finishes*

The application of modular finishes is clearly more complicated than that of trowelled finishes, in that comparatively small modules have to be laid individually to large areas while maintaining an accurate level between each module, and to the floor as a whole. This cannot be carried out even by the most expert tradesmen without permitting a degree of variation. British Standard CP 202 suggests that a tolerance of ± 15 mm compared with the specified datum level could be acceptable and that for localized variations the limit should be ± 3 mm under a 3 m long straight edge. Those figures should be achievable and are a good guide, but, as CP 202 also states, it is important that there should be no appreciable difference in level across individual joints (and that includes movement joints), otherwise damage will occur.

Setting out of tiles dry, or the use of a gauge rod, generally ensures that the tiles are fitted in such a way as to avoid small or unsightly cuts. Cuts must always be made with a special cutter or brick saw. Cropping of bricks or tiles is to be discouraged.

Procedures for laying vary with the type of system being applied and therefore each will be covered separately.

Bedding in cement mortar. Over a separating layer: The separating layer may be polythene sheet, building paper, bituminous felt or a waterproofing membrane. In all instances, the surfaces on which the separating layer is applied must be smooth, so as not to interfere with movement, and must be swept clean immediately before installation. Joints must be lapped by at least 100 mm, and only by less if a proprietary membrane with contrary instructions is used.

Bedding mixes are generally 1:3 to 1:4 cement: sand mortars, laid at 15 mm to 25 mm thick between battens or stringer courses of tiles, tamped and drawn level. Immediately before application of the tiles, neat cement is dusted on to the surface and lightly trowelled, or a neat cement slurry may be applied.

The tiles are then positioned in such a manner that the joints are 6–10 mm wide and firmly beaten into the surface with a rubber or hide mallet until at the appropriate level. As placing of the tiles proceeds they can be humoured so that the joints are to the appropriate line, conducive with providing an aesthetic joint pattern and absorbing the tolerances of the tile to an extent where they are not evident.

The 'stringer course' mentioned is that practice where a rib or ribs of tiles are laid in advance of the main floor areas to line, falls, and level as indicated by the design. They are sometimes constructed on each side of an expansion joint in an alternative bond to the main floor. The stringer course or rib is then used as a guide for levels during subsequent laying. It is normal practice to allow the stringer course to set before filling-in and in that event, slurrying of the bed of the stringer must be carried out when bedding adjacent tiles to prevent excessive absorption of moisture from the bed. Consolidation of the stringer course is just as important as the main floor area and is often neglected with resulting failure in these areas with subsequent trafficking.

Jointing or grouting should be carried out as soon as practical to avoid contamination of the joint area. Cementitious grouts can be placed as soon as the open jointed tiles are firm enough to support weight. Joints to be resin filled should be left open at least 72 hours to allow water from the bed to evaporate. Those resins sensitive to water contamination may require that this time be increased until an appropriate water content is reached. Cementitious jointing is carried out by application of a 1:1 cement: fine sand, slurry, adequately moist to permit application, but avoiding excessive water which will cause shrinkage.

The inference in utilizing resin jointing is that a greater degree of impermeability is required and therefore more care and attention most be given to the application if that is to be achieved. During the bedding operation the edges of the tiles will have become coated with cement and the joint depth will be inconsistent. Before grouting the joint must first be raked out to a consistent depth of at least 12 mm, or deeper if the specification dictates, and the whole of the joint must then be thoroughly swept with a stiff handbrush to remove cement and dust; a wirebrush makes a good tool in this respect.

Resin jointing can be carried out by hand pointing, gunning or by squeegee. The method of application must be governed by the contractor and it will be dependent on the selection of material. Water tolerant epoxy systems are generally squeegeed, as excess material can be rinsed off with water with minimal concern for staining of the tiles.

Resins such as polyester and furane are generally gun jointed or hand pointed to minimize contamination of tile surfaces and the excess resin is scraped off and the joint then gently wiped with a solvent cloth. If excess solvent is used, the tiles will be stained. It is not uncommon for masking compounds to be applied to tile surfaces where polyester or furane resins are used, and most certainly recommended on anti-slip surfaces which can be clogged with excess resin. Masking compounds are water soluble and can be washed off with warm water when the resin jointing is set. The method of use of masking compound must be strictly controlled to avoid contamination of the internal joint edges.

Bedding in semi-dry screed. The use of semi-dry screeds not only reduces water content of the bedding, hence a reduced shrinkage, but allows falls to be created in the bedding from a flat floor surface, without loss of initial support of the tile as would be the case with normal water contents and thick beds. It is necessary to emphasize that semi-dry beds properly laid are a useful and advantageous method of laying tiles and paviors of all thicknesses, but specific attention must be paid to compaction and water content. To a certain extent the two are related, as a low water content makes the screed incompactable even by mechanical means and a poorly compacted screed is very weak. On the other hand, too high a water content gives a lack of support to the tile. For semi-dry screeds utilizing a 1:4 cement sand mix, the total cement:water ratio for optimum performances is between 0:4 and 0.45. Even then compaction

Figure 9.6

Figure 9.7

INSTALLATION OF SPECIALIST FLOORINGS 97

Figure 9.8

should be seen to be carried out effectively and a good guide in this respect is to place 20% excess loose semi-dry mix between the stringers and to compact down to finished bed level. Figures 9.6–9.11 are photographs of sections through semi-dry screeds of varying initial water content and in an uncompacted and compacted state. In all instances they are from the same batch of fine aggregate concrete dry mix of 1:4 ratio.

Figure 9.6: the screed had 0.3% water:cement ratio uncompacted. The finished core was friable and clearly incompletely hydrated.

Figure 9.7: the water:cement ratio was 0.3%, the screed was compacted by 20% from the original level. Hydration was incomplete and the core was intact but friable and very porous. The full thickness of the core was saturated by immersion in water for only 30 seconds.

Figure 9.8: water:cement ratio was 0.5%, the screed uncompacted. Insufficient compaction made the core weak, but hydration was complete.

Figure 9.9: water:cement ratio was 0.5%. The screed was compacted by 20%, which gave an adequately strong core, which offers a satisfactory level of support for modules without danger of floating.

Figure 9.10: water:cement ratio was 0.7%, the screed uncompacted. Even this level of water content does not self compact.

Figure 9.11: water:cement ratio was 0.7%, the screed compacted, which resulted in the densest screed. However, surface water under compaction was such that considerable joint contamination would have occurred and a tendency to floating was observed.

98 SPECIALIST FLOOR FINISHES FOR CONCRETE

Figure 9.9

Figure 9.10

INSTALLATION OF SPECIALIST FLOORINGS 99

Figure 9.11

Mixing. Mixing of semi-dry screeds is best carried out in horizontal forced action open pan mixers and not in free fall concrete mixers. The latter produces balling of the cement with the consequence of poor ratio of cement:sand, and weak zones. The resulting mix of correct water content will, when squeezed in the hand, cohere and not crumble when the pressure is released.

Laying. The substrate on to which the semi-dry screed is to be laid should be slurried with a 1:1 cement sand mix of creamy consistency to act as a bonding agent and to reduce suction from the semi-dry mix. The semi-dry mix is placed loosely on to the primed surface and trowelled out between the battens to the finish bed thickness plus 20%.

The mix should then be beaten down by float and beam or better still by mechanical vibrator. The compacted screed should be drawn off level with a beam. Should the level be less than required, additional mix should be placed and compacted with a float following light raking of the compacted surface. The finished level of the semi-dry bed should allow for the tile to be beaten in by about 6 mm. When the levels are correct the bed should be slurried with a 1:1 cement:sand slurry of creamy consistency and the tiles are placed and beaten into the slurry with a hide or rubber mallet. Joints should be 6–10 mm wide to allow satisfactory cleaning and pointing, and to permit the tile tolerances to be absorbed. Pointing with cement and resin is carried out in the same manner as previously described.

Fully bedding and jointing in resin cement. Most certainly in chemical duties, this system is the preferred method of laying tiles. It is advantageous in all systems although the cost will be higher. Fully bedding and jointing requires that the screeds be laid to falls first, so that the minimum bed thickness can be used. In the event that a membrane is to be placed between the screed and the finish, priming is not required unless the membrane system demands it.

Where no membrane is used, most resin cements will require primer to be applied either as a bonding agent or, in the case of furane resins, as means of preventing contact with the cementitious substrate which will neutralize the acid catalyst.

Priming systems generally utilize the resin base (except furane) plus hardener or catalyst. The substrate must be clean, dry and free from dust, laitance and contamination. Most primers, except those used for furane, are best laid on wet unless they contain solvents.

The appropriate bedding mortar is mixed in accordance with the manufacturer's instructions and buttered on to the tile or pavior including the edge, and placed firmly on to the wet primer and pressed up against the previously laid course and the adjacent tile. The joint is therefore made under compression, with the excess resin exuding from the joint, carefully struck off with the trowel. The placed module is then supported by a dry tile or pavior to prevent sliding away. Joints in fully bedded and jointed systems are thinner than open jointed at 3–4 mm and therefore some of the aesthetics may need to be sacrificed if tile or pavior tolerances are large.

With this method of application, however, considerable technical benefit is achieved, and the system is more resistant to penetration by liquors. Joints laid in compression will also give better support to tile edges under load.

Mixing resin mortars. Certain resin mortars, notably furane and polyester resins, have a tendency to produce exothermic reactions if left in bulk. For this reason large batches should not be produced, and mixed material should be spread about a tray to a maximum depth of 25 mm or less to reduce heat build-up, particularly in summer months. Old set should be removed regularly to avoid contamination of new mixes.

10 In service maintenance of floors

10.1 General

Not for the first time in this publication, it must be stressed that flooring systems are the most important aspect of any purpose-built industrial building. If a floor is used, the chances are that the floor will also be abused and so bring nearer the time when serious maintenance is essential. At that point end users must be prepared to move plant and equipment, or in certain instances drastically reduce or cease production. The very contemplation of that should, to most end users, be incentive enough to carefully maintain a floor, but this is generally not the case.

Maintenance of a floor is not just a question of cleaning and repairing, although these are two essential elements in prolonging life. A floor can be maintained in service 24 hours a day and even before it can be used, by attention to a few operating points.

Avoid the use of steel wheeled vehicles, however small. Use nylon wheels or, if practical, solid rubber, polyurethane or pneumatic tyres.

Be aware of areas where localized damage may take place. Do not specify that floors must withstand dropping of tools and equipment; try and alleviate the problem or use resilient mats, to protect finishes. Identify these areas to contractors and be prepared to change the specification if prevention of attrition is impossible.

Areas where damage occurs despite precautions should be repaired without delay. The effect and extent of damage caused by chemicals penetrating the surface is not always evident to the eye.

Avoid discharge of liquors direct on to floor surfaces including that from condensate drains which can be particularly aggressive to resin based floors.

Use spreader plates under point loads, and be sure that point loads or spread loads do not act too close to channel edges or movement joints, otherwise serious damage may occur. Where strong chemicals are regularly spilt, floor flushing systems are easily and cheaply installed either as an automatic system or operator controlled. They need only be regularly perforated plastic pipework unobtrusively secured against a skirting or perimeter. Costs are therefore insignificant compared to damage by strong chemicals. These are maintenance points which in effect can be carried out by the end user without interfering with the floor finish and from which long term benefit can be derived.

10.2 Cleaning

Cleaning of floors implies different things in different industries. In some, particularly the hygiene industries, cleaning involves the use of chemicals potentially more aggressive than the normal operating spillages and their products of degradation.

Chemical industries predominately utilize water hosing to dilute spillages and non-aggressive neutralizing agents such as soda ash solutions. In some ways, therefore, the chemical plant is a much more easily defined situation, as far as cleaning is concerned, with less insistence on aesthetics and hygiene, and where heavy duty partially aesthetic finishes can be applied, and where cleaning is essentially a method of aiding speedy discharge to drains of contaminants and of neutralizing spillage.

As already implied, the situation with respect to the hygienc industries, i.e. dairies, breweries, food processing and pharmaceuticals, is entirely different. Firstly the specifications for the flooring tend to be less substantial than for the chemical industry, due mainly to the desire to have light coloured, decorative finishes involving the use of thin tiles, bedded on sand cement screeds, pointed with epoxy resins, or resin based semi-decorative toppings, which have been long proven to be resistant to milk products, beer, blood and their degradation products. However, no longer are the floors cleaned with dilute soap solutions and warm or hot water but with exceedingly complex chemicals. Furthermore, specifications for floors are generally issued with a list of chemicals used for cleaning and descaling of pipes and tanks, all of which are destined to be discharged some time or other on to the floor. In an attempt to increase efficiency in cleaning regimes, other chemicals are added to cleaning chemicals as wetting agents, sequestering agents, sanitizers, not to mention degreasing agents, chelating agents and inhibitors. This situation highlights the fact that cleaning agents, as efficient as they now appear to be in achieving the desired standards of hygiene, could be allowed to become the aggressors in respect to floor finishes and drainage systems. Therefore, having cleaned a floor chemically, one can assist in prolonging its operating life by hosing down with copious amounts of warm water. This operation also reduces the effect of overdosing of cleaning chemicals as may occur from time to time, and which cannot be easily identified as the source of problem it is known to be. Certain cleaning agents when left in contact with floor surfaces, particularly anti-slip finishes, have reacted with tile finishes to form hard deposits which clog the profiles and are not removable except with other harsh materials. Adequate rinsing negates this problem, although in hard water areas even water can leave traces of calcite which can be only be removed with acid type cleaners.

Cleaning of anti-slip surfaces requires more effort than that needed for plain surface floors and by far the best results are achieved by stiff brushes, preferably attached to a mechanical scrubbing device. It should be borne in mind that failure to remove fats and greases from anti-slip surfaces will result in considerable loss of benefit in this respect and loss of aesthetic appearance on lighter coloured floors, as dirt is ingrained into the surface.

10.2.1 Steam cleaning

There was a time when floors and flooring materials were designed around their ability to be steam cleaned. With the reservations already expressed with regard to the development of cleaning chemicals in mind, the need for steam cleaning has diminished and with it has gone one of the prime causes of regular failure of resin toppings and (in less common circumstances) tiled floors.

The use of steam cleaning equipment also led to some abuse, such as the regular occurrence of steam jets left playing on floor areas during lunch breaks and so on. Steam, even at low pressure, can create extreme thermal shock to tiled systems and irreversibly damage resin toppings by softening and blistering as conducted heat expands air in the substrate. Because of the mobile nature of steam hoses, their effect cannot be accommodated by more frequent use of movement joints as are used when fixed items of plant generate excessive heat or hot water spillages.

As a method of cleaning, steam is not even efficient. While it may soften fats and greases, it does not saponify in the same way as cleaning agents, and therefore may lead to blocking of drains as fats resolidify.

In general, the use of mechanical acids combined with judicious use of chemicals and warm water will produce the desired results in achieving satisfactory standards of hygiene and cleanliness, and not reduce the life of the floor.

10.2.2 Cleaning problems

Most detergent cleaning manufacturers offer an advisory service for removal of difficult stains and everyday spillages and end users are well advised to consult with them. Be sure, however, to identify clearly the nature of the floor finish to be cleaned. Cleaning problems are predominately associated with the nature of the floor surface, and it is important not to resort to over-aggressive cleaning methods to compensate for difficult-to-clean surfaces.

One common cleaning problem frequently associated with new installations is staining as a result of residue from the laying operations, or from subsequent operations such as painting or plastering. This problem is more noticeable on dry floors and has a tendency to become much less apparent when the floor is wet. When the floor is dry, white or grey patches appear and sometimes whole floor areas are affected. Dirt appears to be ingrained in the surface even when relatively smooth and this can normally be traced to cement dust or mortar mix from laying operations (in the case of tiles) or dust from building operations, plastering, laying or general builder's spoils settling on to the surface. After settling, the cement or plaster dusts absorb moisture and hydrate in the normal way, adhering to the surface. The fine nature of the dust does not make the problem initially obvious. This type of staining 'disappears' when the surfaces are wetted but even after vigorous scrubbing exercises, always manages to reappear when dry, particularly on anti-slip finishes which attract it to a greater degree. Cement staining can be identified by testing a small area of the surface, preferably in a unobtrusive position by applying a small quantity of 5% hydrochloric acid or proprietary brick cleaner and observing for effervescence. Scrub the area with a stiff brush, rinse away with clean water and allow to dry.

If the small test area is noticeably cleaner than the surrounding area, then the problem has been identified and can be dealt with in the manner of the test. Heavily stained areas many need repeat treatment or longer dwell times.

Note: Consult the contractors responsible for laying the floor and ask them to clean the floor first, particularly if they are responsible for the contamination. Check that the floor finish and drains will resist the dilute acid treatment. Ensure protective clothing and face and eye protection is used when working with chemical solutions and allocate responsible personnel to the task.

Stains resulting from the jointing processes are clearly the contractor's responsibility and end users should not attempt to remove these. Dark resins such as furane are notorious for discolouring light coloured tiles and removal is only possible by use of solutions which can have a detrimental effect on the joints themselves. For this reason contractors must be involved, as early as possible when the problem is noticed.

As thin films are much slower curing than joints, successful cleaning can be achieved if early attention is given.

11 Fault finding, problem analysis and solving

11.1 Fault finding

It is generally the case with specialist floor systems that faults find themselves and there are no sure ways of testing all the elements of a floor finish before use. It is doubly important therefore that each element of the finish, and that includes the substrate, must be carefully checked as work proceeds. Membranes can be water tested after installation, but this can be counter-productive with easily damaged membranes on large areas, as protection of membranes by the finish is also extremely important.

Problem floors, whether they be associated with substrate, joint or topping are not always obvious when newly laid. Careful inspection of surfaces may reveal small defects which are in need of remedy, but even a trained eye or ear may only spot substantial problems in a few instances. Qualitative assessment is commonly made by tapping to detect hollow areas which may indicate arching with subsequent loss of resistance to impact, separation or bad laying. On the other hand, depending on the floor construction, it may mean nothing at all.

A more positive and repeatable quantitative testing method for impact resistance is the use of the 'Schmidt hammer'. The Schmidt hammer is a spring loaded device which delivers a consistent sharp impact by way of a hardened steel rod, and measures on a scale a percentage rebound calculation which can be applied to known values as a quantitative assessment and forecast of floor suitability, and can identify variations throughout a floor finish.

The British Ceramic Research Association have extensively tested the Schmidt hammer and found the results over a range of finishes to be reproducible and predictable as a means of determining progressive deterioration.

It would be rather difficult with the wide range of flooring materials available to produce a meaningful table of satisfactory figures for each type of installation as thickness of bed, type of membrane, type of resin and so on will all produce different figures. It is possible, however, to arrive at minimum values for a category of floor as follows which the BCRA have determined as follows:

Heavy duty	Minimum rebound reading	60%
Normal duty	Minimum rebound reading	45%
Light duty	Minimum rebound reading	30%

While these figures are intended to be indicative of impact resistance, conclusions can be drawn in other respects by experienced personnel.

The definitions of these levels of duty for the purposes of the above figures only, are:

(a) *Heavy duty*—Those situations where fork lift trucks (or greater loads) are in use. Typically these would be chemical plants, dairies, breweries and similar installations, but also including (from experience) supermarkets.
(b) *Medium duty*—Pedestrian traffic and light hand trucks only
(c) *Light duty*—Domestic use or equivalent.

It will be seen from this that the majority of installations this publication is intended to cover should therefore have readings of 60 or over unless for pedestrian use only.

To summarize, faults generally exhibit themselves and there is no substitute for adequate supervision during installation. Schmidt hammers, for instance, will detect local variations, arching and poor construction, but there is no other non-destructive method by which the total system can be assured other than use.

11.2 Problem analysis and solving—tiles

11.2.1 *Arching*

Arching at its worst can be clearly seen in floor finishes. It can occur in screeds or in toppings, tiles or in combinations. It need not be visible to be problematic; even 1 mm may lead to localized flexing of the finish with subsequent failure. Arching may be identified visually or by tapping for hollowness and ultimately leads to breakdown of the finish under load in heavy traffic areas.

Causes are poorly constructed or absent movement joints not accommodating expansion from temperature, permanent growth, poor adhesion, or by screeds shrinking and placing tiles in compression.

Remedy: Reconstruct movement joints correctly and replace damaged areas. For tiles arched due to screed shrinkage, removal and replacement of the tiles is the only remedy.

11.2.2 *Tile edge failure*

Tile edge failure is cracking of the tile edges, mainly parallel, the fracture line being from the top face to the edge of the tile. It is symptomatic of edge compression due to inadequate expansion allowances, arching, curling, shrinkage of bed or flexing of the substrate.

Remedy: Remove and replace defective tiles, correct any defective joints and screed.

FAULT FINDING, PROBLEM ANALYSIS AND SOLVING

11.2.3 *Tile separation*

Separation of tiles from the bed is generally the result of poor adhesion, particularly when the failure is clean from the tile. Contributory factors in precipitating the failure can be shrinkage of the bed, expansion, or vibration due to plant or traffic.

Remedy: Remove loose tiles, check those remaining and replace as necessary.

11.2.4 *Tiles breaking up*

Disintegration of tiles is predominantly the result of mechanical damage or excessive trafficking. Tiles with spalled edges, loose tiles, or loose tiles and bed, are more susceptible to damage than properly fixed tiles, therefore under these circumstances the effects of trafficking may be exaggerated and not solely responsible for the failure. Breaking up of tile finishes may also be a result of failure of the bedding system due to over-trafficking, early trafficking, poor compaction, incorrect sand: cement ratios or poor mixing of specialist materials, and the only remedy for these situations is complete removal and replacement. Where this occurs at stringer courses or ribs, poor compaction of the bedding in this area due to the method of construction and lack of attention to compacting procedures is invariably to blame.

11.2.5 *Tile staining*

Staining of tiles when new is generally indicative of bedding cement or jointing cement deposits or smears on the tile surface. The following remedies work because they attack the deposit or stain. Where the stain is caused by jointing material it is logical therefore that the joint itself may be attacked. Seek specialist advice where substantial staining is involved.

White/grey cement stains are best removed with a dilute solution (5%) of hydrochloric acid (provided joints are not cement based also) applied by brush (if general seek specialist advice on handling and application of larger quantities) followed by washing and scrubbing with water and mild detergents. Stubborn areas should be retreated.

Epoxy resin or polyester resin stains must be dealt with immediately. Thin smears of resin will not achieve the same state of cure as the joints in a given time and therefore controlled application of paint strippers such as modified methylene chlorides will soften smears quickly with marginal effect on the joints. If possible, the solvents should be confined to the tiles themselves which are unaffected by prolonged application.

Immediately the stain is loosened or wrinkles, the excess material should be absorbed by sawdust or absorbent powder and removed. Do not flush down drains. Handling of solvents, even non-inflammable solvents should be carried out in well ventilated areas under proper control by competent personnel.

Furane stains. Because of its dark brown to black colour, furane resin is prone to stain tiles particularly those not masked, of light colour or with a profiled surface. Provided the joint material is properly set, the use of dilute bleach solution will remove recent furane stains which will not have properly cured. If the stains have been present for any length of time, the use of concentrated hypochlorite may be the only solution, with attendant problems of potential joint damage and the need for competent personnel to handle the chemical. Strong solutions should be confined to application on the tiles themselves.

11.3 Problem analysis and solving—joints

Joints do not have the comprehensive chemical resistance of the tiles themselves and therefore their selection for a specific duty is all important. Failure to correctly select the jointing material will result in loss of jointing material with consequential attack on the bedding material (if dissimilar or cementitious) loss of support for tile edges, and generally complete failure of the tiling system. Exactly the same problem could arise if the duty is changed from that designated for the original specification without reference to the original supplier of the jointing system.

A common occurrence in jointing systems with cementitious beds, is a failure to properly rake out the bedding to a satisfactory depth. This exhibits itself as a loss of jointing material in isolated areas with bedding exposed or with hollow areas beneath jointing, where chemical attack has taken place. In the latter instances, jointing material commonly collapses into the void.

Localized failures can occur adjacent to condensation traps, vessel discharges and concentrated chemical spillage.

Remedy for any of the above is at the very least raking out and repointing. However, from experience it is generally necessary to remove tiles in affected areas and replace and repoint to enable a check on the substrate or bedding to be made.

11.4 Problem analysis and solving—monolithic finishes

Problem

Arching of substrate screeds derives from the same problems as listed in the section on tiles. While separation of resin or modified toppings from substrates with subsequent arching is not unknown, this problem is invariably related to adhesion failure and in these situations complete replacement is essential.

Problem

The most common causes of blistering in new monolithic finishes occur if moisture from the substrate or solvents from primer systems are trapped during laying procedures. Less frequently, direct sunlight shining on newly laid floors through roof lights, windows or doorways will produce blistering.

Remedy: Blisters cannot be left in monolithic finishes as they are prone to breakdown by traffic with subsequent attack on the substrate leading to larger areas requiring repair. Isolated blisters should be repaired individually, areas containing large blisters should be replaced completely. Unfortunately repairs to completed monolithic flooring systems are all too obvious, but blistering identified before sealing procedures may well not be visible after repair and sealing.

Problem

Surface crazing is more common in cementitious monolithics than in resin toppings, which points the way to evaporation or more correctly rate of evaporation of water from the matrix being responsible. This is also borne out by the fact that floors laid in summer months, dehumidified atmospheres or heated areas are more prone to surface crazing. This form of crazing is particularly noticeable after wetting of the floor surfaces, as the cracks are highlighted due to water retention as as the main floor area dries. This crazing in cementitious monolithics toppings, though resembling surface crazing in concretes is in fact rather more, and generally extends through to the substrate. In its early forms, crack widths are so small as to be difficult to measure but are typically 0.1 mm which, while appearing to be insignificant, is a total loss of integrity and therefore serious. The crack widths develop with trafficking which spalls the edges.

Investigations into this phenomenon in situations which do not meet the normal criteria for occurrence in toppings, have in fact indicated that this form of crazing can be reflective from substrates similarly affected, that is to say that core samples taken through topping and substrate have shown cracks in the substrate corresponding to those in the topping. On the basis that cracking in toppings cannot possibly propagate cracks in the substrate, the inference is clear that the reverse is the case, and therefore great care must be taken in evaluating substrates for suitability. Reflective cracking of this nature is a relatively recently observed problem which it is believed occurs and creates initial stresses in the toppings soon after laying, when cementitious systems are very weak. It is probably precipitated by the change in substrate moisture gradients occurring after laying of the topping, and in the gradual drying taking place in the following months during re-commissioning.

Once the problem has been identified, the environment within the building has generally stabilized; glass strips bonded over cracks away from of traffic areas will generally indicate stability (not be confused with structural cracking), and action need only be taken if the cracking causes problems to processes and so on.

Remedy: Remedies must be considered in the light of the use of the problem floor area. In dry areas there should be little effect on performance. On other areas it may be necessary to overcoat with an elastic floor finish such as polyurethane or, as a last resort remove and replace the defective topping with a resin based floor system with more elasticity.

Because reflective cracking occurs slowly and in the early stages of installation, resin toppings appear less affected as they have the ability to absorb small movements of this nature due to their inherent elasticity in early life, compared with cementitious systems.

Problem

Porosity in monolithic flooring is related to poor formulation (lack of fines), poor compaction or by chemical attack on the resin or filler components.

Remedy: Provided satisfactory adhesion levels can be identified (as porosity can affect these) sealing of the matrix can be a satisfactory solution provided this does not only seal the surface. If adhesion is suspect, replacement must be effected.

11.5 Problem analysis and solving—movement joints

In the sense that they are commonly found to be faulty, movement joints are held to be weak points in floor finishes. They are, however, an essential element and cannot be eliminated except by use of elastic finishes.

11.5.1 *Bridging*

Arched areas of floors between movement joints are a sign of possible bridging of movement joints. The elastic movement joint or suitable packing material must extend to the base of the topping or tiling system, if the space is filled by bedding material or builders spoil the joint becomes bridged and ineffective, placing stresses on the floor.

Remedy: Bridged joints can be cut out to full depth and replaced. However, consequential damage to the adjacent tiling must also be repaired.

11.5.2 *Splitting or adhesion failure*

Lack of adequate elasticity under shrinkage forces has been known to split adequately bonded jointing materials or separate then from the sides if adhesion is not complete. It may occur during relaxation after joints have been under compression for some time on existing floors, under shrinkage forces on new floors if joint specification or application is incorrect or as a result of flexing on suspended floors.

Remedy: Cut out and replace joints to correct specification including any consequential flooring damage.

11.5.3 General problems

Pooling. Pooling of liquor on specialist finishes may be solely due to inadequate falls, lipping (in tile finishes), absence of drainage points or by inadequate positioning of plinths and so on. It may also be, in existing floors where previously not occurring, symptomatic of arching as a result of other problems previously discussed.

12 Case histories

For many reasons, not least the fact that certain problems relating to these case histories may still be *sub judice* or ongoing, the instances quoted here are anonymous but genuine examples of lack of forethought, lack of money, poor specification, poor contracting or a combination of every conceivable 'don't' in this publication. They are offered as the results of situations the readers may well find themselves in. Do not make the mistake of believing that such blatant errors must happen very infrequently.

12.1 Figures 12.1–12.11

Specification 20 mm thick tiles were fully bedded and jointed in furane resin cement direct to polyisobutylene membrane. It is often difficult to demonstrate arching of a floor photographically. You will see we have no problem in that respect here in Figure 12.1.

The problems with this specification are manyfold, the main ones being:

Figure 12.1 Arching of floor or tiles 20 mm thick, bedded and jointed in furane resin cement direct to polyisobutylene membrane

CASE HISTORIES 113

Figure 12.2 Floor collapse after arching

(a) 20 mm thickness is inadequate to lay over loose type membranes; they have insufficient weight and resistance to arching
(b) expansion joints were not fitted to the perimeter; this precipitates arching
(c) the membrane can be clearly seen to be swollen, the spillages contained solvents extremely aggressive to the membrane.

The likely sequence of events was arching due to lack of expansion joints, hairline cracking of the joints followed by ingress of liquors, and permanent contact of the liquor with the membrane with little or no possibility of dilution or drainage. In the long term, the membrane would have failed anyway as it is wrongly specified, but here we have a situation where the client has spent considerable money on floor finishes, utilized membranes directly under the finish as per good technical practice, but ended up with a problem floor because the combination of the right materials, the right application and the right installation are the key to success, not just one of those elements. In Figure 12.2, another part of the same area can be seen collapsed due to traffic, a logical consequence of arching.

Figures 12.3–12.5 are of channel areas. Figure 12.3 shows extreme distress of the finish either side of the channel area as a result of constant removal and replacement of the large steel grids. The breakdown has been accelerated by expansion and the fact that the system has inadequate impact resistance and is of a floating nature.

Figure 12.4 is a detail in an area of heavier chemical spillage. Metal bearer

114 SPECIALIST FLOOR FINISHES FOR CONCRETE

Figure 12.3 Breakdown of finish either side of channel area

Figure 12.4 Channel area breakdown where subject to heavy chemical spillage

Figure 12.5 Channel tiling collapsed, substrate corroded

bars were utilized in the specification and the membrane was terminated at the metal bar. No attempt was made to seal the edge; even so, as can be clearly seen, this would have not prevented breakdown. Once the bearer bar was corroded, simple paths existed for the corrosive liquor to pass behind the channel tiling which can be seen being forced away.

Figure 12.5 shows an adjacent area in an advanced stage of deterioration; channel tiling has collapsed away and the main substrate is badly corroded.

There is no doubt that in the foregoing instance the incorrect specification was chosen. The specification which was used did not comply with basic trade practices and the resulting problems were extremely serious not least for the client whose production was severely affected.

Figure 12.6 shows a plinth in an existing installation. The plinth tiling has been carried out thoughtfully and carefully. Although movement joint has not been included to the perimeter in this instance, at least at the time the photograph was taken this omission had not created any failure. The problem here is lack of housekeeping. The centre of the plinth seen in the middle of the photograph is full of corrosive liquor which has accumulated over a period of time and not been neutralized, creating a deeper and deeper cavity. Support for the pump is virtually non-existent and the pump in position is clearly not that originally intended for this plinth indicating a hasty maintenance change which has turned into a permanent fixture with predictable results.

Figure 12.7 is a further example of poor housekeeping in a similar situation. A

Figure 12.6 Plinth tiling: corrosive liquor can accumulate—if not neutralised, can produce a cavity

Figure 12.7 Leaking pump near plinth can cause damage if run continually, especially if floor corrosion already apparent

Figure 12.8 Damage to plinth tiling due to leaking pump, at same plinth as in Figure 12.7

badly leaking pump is run continually even though severe corrosion has taken place.

Figure 12.8 is the same plinth viewed from the opposite side. Only the tile wedge (taken from the floor) supports the end of the pump frame.

In these instances the tops of the plinths were not sealed against ingress of liquor or skirted by movement joints and subsequent problems have not been mitigated by the end user's failure to maintain equipment or have remedial works carried out.

Figure 12.9 shows a specification failure. Lead is currently an uncommon flooring membrane but in this sulphuric acid area it has been utilized as a tanking beneath paviors and as an exposed protection to plinths. Beneath the pavior protection the lead was perfectly sound, but exposed areas were subject to chemical attack due to the fact that the grade of lead used was inferior and not a chemical resistant grade. Pump leakages attacked the pump covering and ultimately liquor would have access to the main substrate. Tiling the plinth in addition to the lead would probably have sufficed, although over the long term, deterioration of the lead would have been inevitable.

Figures 12.10 and 12.11 show the ultimate effects of failure to identify and resolve problems which have arisen related to inadequate protection to floor areas.

Figures 12.12 and 12.13 are of a swimming pool surround. Figure 12.12 should

Figure 12.9 Failure of lead tanking after contact with sulphuric acid owing to poor specification of lead

depict movement joints either side of the small tiles (as these were added as part of a planned maintenance). However, a cement infill can clearly be seen beneath the movement jointing material, which has been applied as a thin coating instead of full depth. The ultimate consequence of this is that the stainless steel bearer bar for the channel cover has been distorted by expansion forces (Figure 12.13) and the grating will become increasingly difficult to remove.

One of the major causes of problems in cement bedded resin jointed systems is failure to rake out for the resin jointing to an adequate depth and the resin jointing can easily break away or be permeated with consequential attack on the cementitious material.

Figures 12.14–12.17 show various sections through a resin jointed system. Figure 12.14 shows a void beneath the joint. The resin depth is about 2 mm. Figure 12.15 shows cement beneath the joint but the joint is very shallow.

CASE HISTORIES 119

Figure 12.10 Failure after inadequate protection of floor

Figure 12.11 More failure as a result of inadequate floor protection

SPECIALIST FLOOR FINISHES FOR CONCRETE

Figure 12.12 Swimming pool surround: cement infill applied beneath movement jointing material, as thin coating instead of the full depth

Figure 12.13 Swimming pool surround: effect of poor cement infill shown in Figure 12.12; steel bar distorted by expansion forces

CASE HISTORIES

Figure 12.14 Resin jointed system: void beneath joint

Figure 12.15 Resin jointed system: shallow joint with cement beneath

Figure 12.16 Insufficient depth raked out before joint was formed

Figure 12.17 Resin jointed system: collapse of materials after inadequate support

CASE HISTORIES 123

Figure 12.16 is a clear example of an application where inadequate depth has been raked out, and what cannot be transmitted by the photograph is the fact that the cementitious jointing is uncompacted and very friable. This occurs because of lack of compaction of the brushed in mix, which deteriorates rapidly even under dilute chemical contact.

In Figure 12.17 all aspects of the problem can be seen, hollowness, shallow joints eroded underneath, and the shallow depth line to which the resin was grouted at the end. A peculiar aspect of this installation was that the paviors were actually bedded on a thin resin bed (which can be seen in the photographs) and then a cement mix was brushed into the joints and topped with resin. Under normal operating conditions, the jointing materials being unsupported or only partly so, collapsed with subsequent ingress of liquors.

The following photographs illustrate the problems which can be encountered due to lack of attention to construction joints and terminating points in substrates to receive monolithic toppings.

Never believe that if the topping cracks over a construction joint that it can be subsequently cut out and filled with an expansion joint; similarly, failure to achieve a firm mechanical key at edges may result in curling.

Figure 12.18 shows a crack 'following' a construction joint. Figure 12.19 is a section through the toppings at that point and local thickening over the original construction joint can be seen. The crack in the topping has occurred some 20 mm away from the site of the joint in the substrate and in a meandering way which makes cutting out in an aesthetic manner impossible.

In Figure 12.20, the terminating chase is hardly formed in such a manner that a mechanical key can be achieved. In this instance, the terminating edge curled away from the substrate and was broken off.

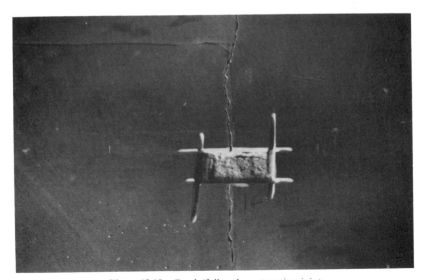

Figure 12.18 Crack 'follows' construction joint

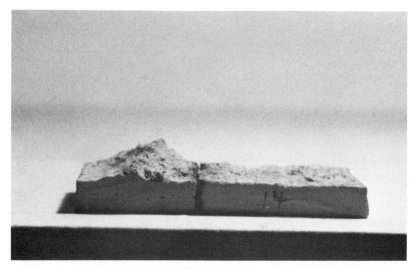

Figure 12.19

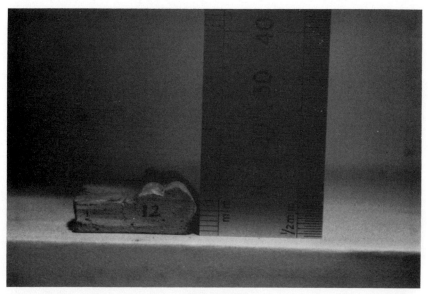

Figure 12.20

In all of these installations, a combination of proper specification or thoughtful and professional application would have saved a considerable amount of financial burden and production downtime and inconvenience.

13 Chemical resistance table

The writer does not favour the use or interpretation of chemical resistance tables by inexperienced personnel without adequate explanation as to the limitations of their use.

13.1 Limitations of use

Because of the wide range of formulation possibilities, hardener systems and catalysts, chemical resistance or otherwise in each of the material categories varies considerably depending supplier and therefore this information is offered as a guide only for floor spillages, and not for containment which requires more comprehensive resistance parameters.

Chemical resistance, while it may be the first consideration in selection of a material, is not the only parameter used to determine the best system and some typical compatibility problems with installations are given below.

(a) Furanes cannot be used direct to concrete as screeds, and the only colour available is black.
(b) Polyesters have obnoxious odours from the styrene content and have tendencies to high shrinkage when used as toppings.
(c) Polyurethanes are sensitive to moisture before cure and are limited in their use as mortars.

Fitness for purpose therefore requires an experienced assessment of the total installation situation which cannot be related in a list of chemicals. Use this list therefore with due warning of this.

SPECIALIST FLOOR FINISHES FOR CONCRETE

Chemical resistance table. *Key:* S—suitable; L—limited; NR—not recommended; (C)—carbon filled grade; FN—furane; EP—epoxy; POL—polyester; PU—polyurethane; LX—latex.

	Conc. %	FN	EP	POL	PU	LX
Acetaldehyde	All	S	S	NR	S	NR
Acetic acid	10	S	NR	S	S	NR
	75	S	NR	S	NR	NR
Acetic anhydride	All	S	S	S	S	NR
Acetone	10	S	NR	S	NR	NR
Alcohol	All	S	S	S	S	S
Alum	All	S	S	S	S	S
Aluminium chloride	All	S	S	S	S	S
Aluminium fluoride	All	S(C)	S(C)	S(C)	S	S
Aluminium nitrate	10	S	S	S	S	NR
Aluminium sulphate	All	S	S	S	S	S
Ammonium acetate	65	S	S	S	S	NR
Ammonium benzoate	All	S	S	S	S	NR
Ammonium bicarbonate	All	S	S	S	S	S
Ammonium carbonate	All	S	S	S	S	S
Ammonium chloride	All	S	S	S	S	S
Ammonium hydroxide	All	S	S	S	S	S
Ammonium nitrate	All	S	S	S	S	NR
Ammonium persulphate	All	S	S	S	S	NR
Ammonium phosphate	22	S	S	S	S	S
Ammonium sulphate	All	S	S	S	S	S
Ammonium thiocyanate	All	S	S	S	S	NR
Amyl acetate	All	S	S	S	L	NR
Amyl chloride	All	S	S	NR	S	NR
Aniline	All	S	NR	NR	NR	NR
Aniline sulphate	All	S	S	S	S	S
Aqua regia	All	NR	NR	NR	NR	NR
Barium carbonate	All	S	S	S	S	S
Barium chloride	All	S	S	S	S	S
Barium hydroxide	All	S	S	S	S	S
Barium sulphate	All	S	S	S	S	S
Barium sulphide	All	S	S	S	S	S
Beer	All	S	S	S	S	S
Benzaldehyde	100	S	S	NR	S	NR
Benzene	All	S	NR	NR	NR	NR
Benzoic acid	All	S	S	S	S	S
Benzyl alcohol	All	S	S	S	S	S
Benzyl chloride	All	S	S	S	S	S
Boric acid	Sat'd (40%)	S	S	S	S	NR
Bromine		NR	NR	NR	NR	NR
Butyl acetate	All	S	S	S	L	S
Butyl alcohol	All	S	S	S	S	S
Butyric acid	All	S	S	S	S	NR
Cadmium chloride	All	S	S	S	S	S
Calcium bisulphite	All	S	S	S	S	S
Calcium carbonate	Dust	S	S	S	S	S
Calcium chlorate	All	S	S	S	S	S
Calcium chloride	All	S	S	S	S	S
Calcium hydroxide	All	S	S	S	S	S
Calcium hypochlorite	20	NR	NR	S	S	NR

CHEMICAL RESISTANCE TABLE

Chemical resistance table (contd.)

	Conc. %	FN	EP	POL	PU	LX
Calcium nitrate	All	S	S	S	S	NR
Calcium sulphate	All	S	S	S	S	S
Carbon disulphide	100	S	S	NR	S	NR
Carbon tetrachloride	100	S	NR	S	NR	NR
Castor oil	100	S	S	S	S	NR
Chlorine gas (wet or dry)		NR	NR	S	NR	NR
Chlorine dioxide	15	NR	NR	S	NR	NR
Chlorine water	Sat'd	NR	NR	S	S	NR
Chloracetic acid	25	S	NR	S	L	NR
	50	S	NR	S	NR	NR
Chloroform	100	S	NR	NR	NR	NR
Chlorosulphonic acid	100	S	NR	NR	NR	NR
Chrome plating solution		NR	NR	NR	S	NR
Chromic acid	5	NR	S	S	S	NR
Citric acid	All	S	S	S	S	NR
Copper chloride	All	S	S	S	S	S
Copper cyanide	All	S	S	S	S	S
Copper sulphate	All	S	S	S	S	S
Cyclohexane	All	S	NR	NR	NR	NR
Detergents sulphonated	All	S	S	S	S	NR
Diethylene glycol	100	S	S	S	S	NR
Dimethyl phthalate	100	S	S	S	S	S
Dioctyl phthalate	100	S	S	S	S	S
Ethyl acetate	100	S	NR	NR	NR	NR
Ethyl alcohol	All	S	S	S	S	S
Ethylene chloride	100	S	NR	NR	NR	NR
Ethylene glycol	All	S	S	S	S	NR
Fatty acids	All	S	S	S	S	NR
Ferric chloride	All	S	S	S	S	S
Ferric nitrate	All	S	S	S	S	S
Ferric sulphate	All	S	S	S	S	S
Ferrous chloride	All	S	S	S	S	S
Ferrous nitrate	All	S	S	S	S	S
Ferrous sulphate	All	S	S	S	S	S
Fluorosilicic acid	10	S(C)	NR	S(C)	S	NR
Formaldehyde solution	10	S	S	S	S	S
Formic acid	10	S	S	S	S	NR
Glycerine	100	S	S	S	S	NR
Gold plating solution		S	S	S	S	NR
Hydrobromic acid	18	S	NR	S	S	NR
	48	NR	NR	S	NR	NR
Hydrochloric acid	All	S	S	S	S	NR
Hydrofluoric acid	All	S(C)	NR	S(C)	NR	NR
Hydrofluorosilicic acid	All	S(C)	NR	S(C)	NR	NR
Hydrogen peroxide	5	NR	S	S	NR	NR
	30	NR	NR	S	NR	NR
Isopropyl alcohol	All	S	S	S	NR	S
Jet fuel (JP-4)		S	S	S	S	NR
Lactic acid	All	S	S	S	S	NR
Linseed oil	100	S	S	S	S	NR

Chemical resistance table (contd.)

	Conc. %	FN	EP	POL	PU	LX
Magnesium carbonate	20	S	S	S	S	S
Magnesium chloride	All	S	S	S	S	S
Magnesium sulphate	All	S	S	S	S	S
Maleic acid	40	S	S	S	S	NR
Methyl alcohol	All	S	S	S	NR	NR
Methylene chloride		NR	NR	NR	NR	NR
Methylene ethyl ketone	100	S	NR	NR	NR	NR
Milk and milk products	All	S	S	S	S	S
Monochlorobenzene	100	S	NR	NR	NR	NR
Naphtha	100	S	S	S	L	NR
Nickel chloride	All	S	S	S	S	S
Nickel nitrate	All	S	S	S	S	S
Nickel plating solution		S	S	S	S	NR
Nickel sulphate	All	S	S	S	S	S
Nitric acid	5	NR	S	S	S	NR
	30	NR	NR	S	L	NR
	60	NR	NR	NR	NR	NR
	Fumes	NR	S	S	NR	NR
Nitric/chromic	15/3	NR	NR	NR	NR	NR
Nitric/hydrofluoric	18/4	NR	NR	S(C)	NR	NR
Nitrobenzene	5	S	NR	S	NR	NR
Oleum		NR	NR	NR	NR	NR
Oxalic acid	Sat'd	S	S	S	S	NR
Paraffin (kerosene)	100	S	S	S	S	NR
Petrol		S	S	S	S	NR
Phenol	10	S	NR	NR	S	NR
Phosphoric acid	80	S(C)	S(C)	S(C)	S	NR
Phthalic anhydride	Sat'd	S	S	S	S	S
Plating solutions						
Cadmium cyanide		S	S	S	S	NR
Chrome		NR	NR	NR	S	NR
Gold		S	S	S	S	NR
Lead		S	S	S	S	NR
Nickel		S	S	S	S	NR
Platinum		S	S	S	S	NR
Silver		S	S	S	S	NR
Tin fluoroborate		S(C)	NR	S(C)	S	NR
Zinc fluoroborate		S(C)	NR	S(C)	S	NR
Polyphosphoric acid		S(C)	NR	S(C)	S	NR
Polyvinylacetate emulsion		S	S	S	S	S
Potassium carbonate	10	S	S	S	S	S
	50	S	S	S	S	NR
Potassium chloride	All	S	S	S	S	S
Potassium dichromate	All	S	S	S	NR	NR
Potassium hydroxide	10	S	S	S	S	S
Potassium permanganate	Sat'd	S	S	S	S	NR
Potassium sulphate	All	S	S	S	S	NR
Propylene glycol	All	S	S	S	S	NR
Pyridine	100	S	NR	NR	NR	NR
Silver nitrate	All	S	S	S	S	NR
Silver plating solution		S	S	S	S	NR
Sodium bicarbonate	10	S	S	S	S	S

CHEMICAL RESISTANCE TABLE

Chemical resistance table (contd.)

	Conc. %	FN	EP	POL	PU	LX
Sodium bisulphate	All	S	S	S	S	NR
Sodium carbonate	10	S	S	S	S	S
	35	S	S	S	S	NR
Sodium chlorate	50	S	S	S	S	NR
Sodium chloride	All	S	S	S	S	S
Sodium chlorite	All	S	S	S	S	NR
Sodium chromate	50	S	S	S	S	NR
Sodium cyanide	All	S	S	S	S	S
Sodium ferricyanide	All	S	S	S	S	NR
Sodium hydroxide	5	S	S	S	S	S
	25	S	S	S	S	NR
	50	S(C)	NR	NR	S	NR
Sodium hypochlorite	15	NR	NR	S	S	NR
Sodium silicate	All	S	S	S	S	S
Sodium sulphate	All	S	S	S	S	S
Sodium sulphide	All	S	S	S	S	S
Sodium sulphite	All	S	S	S	S	S
Stearic acid	All	S	S	S	S	NR
Styrene	100	S	S	NR	S	NR
Sugar/sucrose	All	S	S	S	S	S
Sulphated detergents	100	S	S	S	S	NR
Sulphur dioxide (dry or wet)		S	S	S	S	NR
Sulphur trioxide		S	S	S	S	NR
Sulphuric acid	10	S	S	S	S	S
	25	S	S	S	S	NR
	50	S	NR	S	S	NR
	70	S	NR	S	L	NR
	75	S	NR	S	L	NR
	Fumes	S	S	S	S	S
Superphosphoric acid		S(C)	NR	S(C)	S	NR
Tannic acid	All	S	S	S	S	NR
Tartaric acid	All	S	S	S	S	NR
Toluene	100	S	NR	NR	NR	NR
Transformer oils		S	S	S	S	NR
Trichloroacetic acid	50	S	NR	S	NR	NR
Trichloroethylene	100	S	NR	NR	NR	NR
Vinegar		S	S	S	S	S
Water, distilled		S	S	S	S	S
Xylene	100	S	NR	S	NR	NR
Zinc chloride	All	S	S	S	S	S
Zinc fluoroborate (plating)		S(C)	S(C)	S(C)	S	NR
Zinc sulphate	All	S	S	S	S	S

Information with respect to chemical resistance given above is based on generalizations for the materials listed. It is given in good faith, and no warranty is given or implied for a specific product.

Index

abattoirs 7
accelerators in resin cross-linking 28
acid etching 55
acid resistance 25, 26
    (*see also* chemical resistance)
    of furane resin mortars 31
    of polyester resin mortars 30
adhesion 13, 24
aggregates
    abrasion-resistant 38
    formulation of 8
    heavy-duty 63
aggressive environments 28, 35
air inhibition 10
alkali resistance 25, 26
    (*see also* chemical resistance)
alloy angles 80
amines (*see also* hardener systems)
    aromatic 26
    cycloaliphatic 24, 26
    polyamido- 24, 26
    primary and tertiary 23, 24
anti-slip coatings 19
anti-slip properties 18, 20
anti-slip surfaces 7, 8, 11
anti-slip tile finishes 19, 20, 21
anti-static systems 38, 39
arching 106, 112–13
asphaltic membranes 44
atmospheric moisture *see* moisture

barrier cream 26, 30
bauxite in fillers 38
bearer bars 50
bed thickness 69–70
bedding
    in cement mortar 94–5
    materials for 67
    in resin cement 99–100
    in semi-dry screed 95–9
    systems for 16
bedding mixes for tiles 94
benzoyl peroxide 28
bisphenol resins 28
bitumen
    as membrane 41–2
    in paints 17
bitumen-impregnated cork 81

bituminous mastics in movement
    joints 81, 82
bleaches, resistance to 28
break-up, tile 107
breweries 7, 17, 27, 28, 34, 102
bridging of joints 80, 110

calculation parameters for movement
    joints 83
capillaries 88
case histories 112–24
catalysis
    of furane resins 31
    of polyester resins 28
    of polyurethane resins 33
catalyst systems 22
    and fillers 36
catalysts
    latent 31
    in resin cross-linking 28–9
cement
    acid-resistant 13
    high-alumina 34
cement staining 104
cement-to-sand ratio 62
channel drainage, types of 47–50
channel gratings 47
channels 72
    membranes in 72–3
chemical properties
    see under individual systems
chemical resistance 12, 14, 21, 24–6
    table of 125–9
cleaning
    ease of 18
    in hygiene industries 102
    procedures for 102–4
    problems in 103–4
clothing, protective 21, 30
coatings, epoxy resin 22
colour, stability of 24
columns 50, 53, 54, 71
compressive strength
    see strength
concrete
    preparation of 4
    wet, effect of 57
construction joints *see* joints

INDEX

contamination
 by chemicals 58−9
 by oils and greases 57−8
 removal of 57−9
 of site 85
copolymers
 acrylic 5
 natural rubber 5
 styrene butadiene 5
 vinylidene chloride 5
corundum in fillers 38−9
cracking 14, 106, 109
crazing 109
cross-linking in polyester resins 29, 30
curing 24, 91
cycloaliphatic amines see amines
cyclohexanone peroxide 28

dairy industry 7, 17, 27, 28, 34, 102
diluents, reactive and non-reactive 23
diphenylmethane-4-4′-diisocyanate (MDI) 32, 33
drainage 46−50

edge failure, tile 106
end user, maintenance by 20
engineering brick 7
epoxies/polysulphides, modified 81
epoxy 7
epoxy equivalent weight 23
epoxy resin 22
 water-miscible 5
epoxy resin systems
 aesthetic appearance of 27
 chemical properties of 24
 compared to polyurethane resins 32
 hardener systems for 23, 24, 26
 moisture tolerance of 24
 physical properties of 25
 and replacement fillers 38
 resilience of 24
 as skin irritants 24
 as toppings 27
 used in flooring 26
epoxy terrazzo 93
exothermic reaction 29
 in fillers 36
 in furane resins 31
 in polyester 10
 in resins 16
expansion 10
 coefficients of 6, 32, 83
 expansion joints see joints

fall 46, 47
 degree of 46, 48, 49
 and thin finishes 46, 51
 with screeds 50−1

fibreboard 81
filler ratio 37
fillers 36−9
 bulk 38
 in epoxy resin systems 22
 extending 36
 properties of 36
 siliceous 38
 for specialist finishes 38
 specialty 38
 in toppings 37
 uses of 36
financial considerations 3
finishes (see also modular finishes)
 application of 90−100
 design of 61−76
 polymer modified cementitious 90−1
 removal of existing 56−7
fixing systems, design of 66−7
flexing 14
flexural strength see strength
flint, calcined, in fillers 39
floating membranes see membranes
floor finishes 4
 modular 5
 paint and sealers 4
 polymer or resin modified cementitious 5
 resin 7
flow properties 92
food processing industries 10, 17, 23, 26, 28, 33, 35, 102
freedom from interference on site 85
furane resins 12, 31−2
 acid resistance of 31
 and arching 112
 chemical properties of 31
 cohesive strength of 31
 jointing in 95
 physical properties of 32
 solvent resistance of 31
 and tile staining 108

granite in fillers 39
graphite in fillers 38
grouts, cementitious 95
gulley outlets 48, 49, 75

hardener systems 22, 23
 amines in 23, 24, 26
 chemical properties of 24
heat distortion 25
heavy-duty aggregate 63
housekeeping 115−16
humidity, effect of in polyurethane resins 33
hygiene, personal 30
hygiene industries 13, 16, 27, 33, 102

131

INDEX

impermeability of membranes 40
impervious membranes *see* membranes
installation 86–100
installation design 63–5, 68–76
isocyanates
　as catalytic modifiers 29
　in curing 11
isophthalic resins 28

joint fillers 81
joints
　board 54
　construction 92, 123
　expansion 77, 113, 123
　movement 45–6, 52–3, 65, 66,
　　77–83, 110–11
　　calculation parameters for 83
　　construction of 79
　　function of 77
　　installation of 82
　　materials for 81
　　position of 78
　　problems in 110–11
　perimeter 52
　problems in 108, 110–11

latex cements and mortars 34–5
　properties of 35
latex foam in fillers 81
latex screeds 21
latices, types of 7
laying in bond 68
liquors, discharge of 101
loading characteristics 3
luting flanges 41

maintenance of floors 20, 101-4
masking compound 20
MDI
　　see diphenylmethane-4-4′-diisocyanate
MEK peroxide 28
membrane sheeting 86-7
membranes 12-14, 40-4, 66, 86–90
　application of 86–90
　floating 40
　function of 40-1
　impervious 7
　liquid 89
　need for 40
　polyurethane resin 34
　pressure-sensitivity of 90
　termination of 88-9
　types of 41-4
modified cementitious systems
　6
　epoxypolysulphide 81
　polymer 7
　resin 7

modular finishes 12, 65, 68, 93
　application of 93-4
　design of 65-76
moisture
　atmospheric, and polyester resins 29
　effect of in substrates 57
moisture contamination, effects of 24
moisture content of substrate 46
moisture curing 11
moisture tolerance 24, 27, 29
molecular sieves in fillers 38
monolithic finishing, problems in 108-10
mortars
　chemically-resistant 27, 28, 30, 34, 38
　epoxy resin 17, 26
　furane resin 17
　latex cement 34-5
　polyester resin 17, 30
　polyurethane resin 34
　silicate 17

odour 28
　from installation 22
　from solvents 11
　from styrene 10
outlet spigots 76
outlets, membrane laying and 87-8
　(*see also* gulley outlets)
overspecification 8
oxidizing chemicals 9, 28

paint finishes 4
　removal of 56, 58
pavior finishes 56
pavior systems 12, 13, 15
perimeter upstand 53
permanent growth 77
peroxide catalysts 28
pharmaceutical industry 7, 17, 102
physical properties
　see under individual systems
plastics as membranes 43
plinths 50, 53, 71-2
　design of 53-4
point loads on membranes 90
polyamidoamines 24, 26
polyester 7
　properties of 10
polyester resins
　catalysts for 28
　chemical properties of 30
　composition of 28
　curing rate of 29
　disadvantages of 28
　formulation of 28
　jointing in 95
　moisture tolerance of 29
　physical properties of 30

INDEX

as skin irritants 30
as toppings 31
types of 28
use of with replacement fillers 38
uses of 30
polyethylene as membrane 41, 43-4
polyisobutylene (PIB) as membrane 43, 112
polyols as resin base 32, 33
polypropylene as membrane 43-4
polystyrene as joint filler 81
polysulphides in movement joints 81
polythene 14
polyurethane 7, 10, 14
  as membrane 42
  in movement joints 81
polyurethane foam as joint filler 81
polyurethane resins 32-4
  catalytic formation of 32, 33
  chemical properties of 33-4
  curing rate of 33
  physical properties of 32
  uses of 34
pot life 29
powder phase 36-9 (see also fillers)
primers
  epoxy resin 22
  water-dispersible 24
priming 92
problem analysis 106-11
product design 65-8
protection of finishes on site 85
protuberances 52
pulverized fly ash in fillers 39

rate of cure 29
rebates, channel 73
release oils 45
resilience 24
resin beds, thin 46
resin cement, bedding in 99-100
resin hardener ratio 23
resin jointing 95
resin mortars
  filler ratio for 37
  mixing of 100
  plasticity of 37
resin-removing creams 26, 30
resin systems (see also individual systems)
  epoxy 22-7
  furane 31-2
  physical properties of 25
  polyester 27-31
  polyurethane 32-4
  and replacement fillers 38
resin toppings, fillers and 36-8
resins (see also under individual systems)
  chemical resistance of 8

choice of properties of 8
diluented 9
epoxy 8-10
furane 8, 12
physical properties of 8
polyester 8, 10
polyurethane 8, 10
sensitivity to moisture of 11
synthetic, for toppings 7, 22-35
rubber (see also latex)
  silicone, in movement joints 81
  sheet, for flooring 21
  for membranes 43
  vulcanized 21

safety clothing 21, 30
sands, properties of 36-7
scabbling 55
Schmidt hammer 105
screeds
  bonded 51
  curling of 51
  design of 61-3
  isolated 50
  methods of placing of 50-1
  polymer modified cementitious 61-2
  semi-dry 13, 67, 95-9
sealers 4, 22
self-adhesive aluminium type 44
self-levelling (self-smoothing) systems 9, 23, 92
semi-dry beds see screeds
separation, tile 107
service ducts 50
sheet rubber see rubber
shrinkage 10, 28
silicates in sealers 5
silicone rubber see rubber
site
  contamination of 85
  design considerations of 66
  requirements for 84-5
  temperature of 84
skin, protection of 26, 30
slipperiness 20 (see also anti-slip)
slump 16
solvent resistance 25, 28, 29, 31
solvent taint 4, 10 (see also odour)
solvent vapours 4
solvents, degreasing 58
specification, choice of 3, 8
staining 27, 107-8
steam cleaning 103
steel, corroded 60
steel floors see substrates
steel-wheeled vehicles 101
stopper courses 70
storage of materials 85

strength (of resin systems)
  compressive 25, 32, 35
  flexural 25, 32, 35
  tensile 32, 35
stringer course 94
styrene monomer 28, 29
substrates 13, 16
  concrete 40, 55
  design of 45-60
  drainage system for 46
  existing 45, 57, 59, 60
  materials for 59-60
  moisture content of 24, 46
  movement joints in 52
  new 45, 55
  non-cementatious 54
  plinths in 53
  preparation of 11, 55
  repair of 59-60
  specialist finishes for 54
  steel 40, 54, 60
  surface finish of 46, 51-2
  timber 54
surface crazing 109
surface deposits 20
surface finish 46, 51-2
synthetic resin membranes 42

tank linings 12
temperature gradients 83
temperature, effect of in movement joints 77
tensile strength see strength
terminating batten
threshold of smell 28
tile finishes
  degree of fall for 46
  problems in 106-8
tile floors
  bedding/jointing systems for 16

elastically bonded 14
floating 13, 14
full bonded 13
tile joints 67
tile specification 12
tiles
  choice of 15
  laying of on membranes 90
  problems with 20, 106-8
  setting out of 94
  staining of 20
  vitrified 15
timber floors see substrates
tolerance on module length 70
toppings
  application of 91-3
  degree of fall in 46
  design of 63-5
  epoxy resin 22, 26, 27
  polyester resin 28, 29, 31
  polyurethane 34
  removal of 56
  resin 62-3, 91-3
  resin modified cementitious 62
  trowelled resin 93
toxicity of hardener systems 24
traffic, movement joints and 78

vacuum shotblasting 55, 58
vibration, movement joints and 77
vibrator, mechanical 99
vinyl ester resin 28
vinylidene chloride 7
viscosity, reduction of 23

water:cement ratio 97
water hosing (waterblasting) 58, 59
waterproofing 40
weather protection 84
weep holes 75, 76